MACMILLAN/McGRAW-HILL

Math

Leveled Problem Solving

Grade 3

To the Teacher

Leveled Problem Solving offers differentiated instruction for all students. Each page provides leveled word problems to be used in conjunction with *Macmillan/McGraw-Hill Math*. The problems correlate to each lesson's objectives and ensure that students can read, solve, and explain their thinking about math problems.

The word problems are written at three levels of difficulty. **Basic** problems (questions 1 and 2) are presented in easy-to-follow language for students who need help with basic skills. **On Level** problems (questions 3 and 4) match the language of the actual lesson for students who have the required skills. **Challenge** problems (questions 5 and 6) offer an opportunity for enrichment for those students who excel.

Leveled Problem Solving serves as a math literacy tool to help all students become successful problem solvers.

Contents

Counting and Number Patterns

Solve.

1. Jan skip-counts by twos. She starts at 2 and ends at 10. Write Jan's numbers.

   ____

2. Greg skip-counts by threes. He starts at 6. He skip-counts three numbers. Which number does Greg end on? ____________

   Is it an odd or even number?

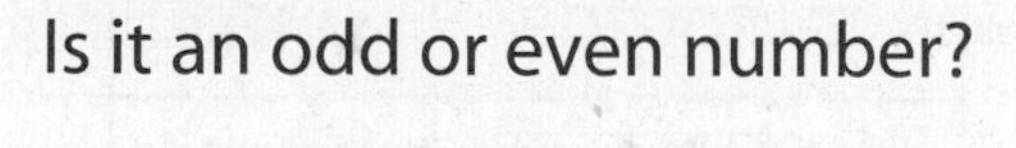

   ____

Solve.

3. Margo lives at 34 Chestnut Street. The house numbers on her side of the street skip-count by twos. There are 5 houses on her street. If Margo's house is the first house, what is the address of the last house?

   ____

4. There are 447 students in Washington Elementary School. There are 398 students in King Elementary School. Which school has an even number of students? How can you tell?

   ____

Solve.

5. Helen draws 27 stars. She colors every fourth star red. How many red stars are there?

   ____

6. Trey started at 12 and skip-counted to 27. What are all the possible numbers he could skip-count by?

   ____

Explore Place Value

Solve.

1. The place-value models show the number of books sold at the School Book Sale last week.

hundreds	ones

How many books were sold?

____

2. Sandra used these place-value models to show how many fish she has in her tank.

tens	ones

How many fish does Sandra have?

____

Solve.

3. Sal weighs 124 pounds. If he wants to show his weight using place–value models, how many hundreds models will he need?

____

4. Tim is 14 years old. How many tens and ones can he use to show his age?

____

Solve.

5. Mel uses 6 place-value models to show the number of video games he has. Two of the models are tens. The rest are ones. How many video games does Mel have?

____

6. Ari has 6 hundreds, 7 tens, and 5 ones in his box of place-value models. He wants to show the number 778. How many more of each model does he need?

____

Place Value Through Thousands

Solve.

1. There are 3,431 seats in a local sports arena. Write that number in expanded form.

____

2. How many tens are in the number 2,840?

____

Solve.

3. The Thompsons traveled 1,572 miles to Florida. How many more hundreds than ones are in the number of miles they traveled?

____

4. Mr. Holt wrote this number on the board:

   5,000 + 30 + 6

   What number is this in standard form?

____

Solve.

5. A science dictionary in the library has 1,634 pages. A Spanish dictionary has 300 more pages than the science dictionary. How many pages are in the Spanish dictionary?

____

6. Kevin is in seat number 1,024. The number on Megan's seat has the same number of thousands and tens as Kevin's number, but 2 more hundreds and 3 fewer ones than Kevin's number. What is Megan's seat number?

____

1.4

Place Value Through Hundred Thousands

Solve.

1. There are 123,456 books in the South Street Library. Write that number in expanded form.

____

2. Ocean City has 231,506 people. What is the value of the 3 in this number?

____

Solve.

3. Last month, Fresh Juice Company sold 140,016 bottles of orange juice. This month, the number of bottles sold was 3 thousand more than last month's number. How many bottles of orange juice were sold this month?

____

4. There are $600{,}000 + 4{,}000 + 800$ seconds in a week. In standard form, how many seconds are in a week?

____

Solve.

5. Solve these number sentences to make a 6-digit number.

$4 + 4 =$ _____ hundred-thousands

$4 - 2 =$ _____ ten-thousands

$8 - 4 =$ _____ thousands

There are no hundreds, tens, or ones.

What is the 6-digit number?

____

6. Use the two clues below to find a 6-digit number.

Clue 1: Each digit increases by one. For example: 456,789.

Clue 2: If you add all 6 numbers, the answer is 21.

What is the 6-digit number?

____

Explore Money

Solve.

1. Sam pays for the apple with a $1 bill. How much change does Sam get back?

____

2. Mi Sook buys a banana. She gets $0.43 change. What coins could the change be?

____

Solve.

3. If you buy a bag of grapes and pay with a $5-bill, how much change will you get back? Tell the bills and coins of the change. Use play money to help.

____

4. Rita buys a bag of oranges. She pays with a $10 bill. Name the bills and coins of the correct change two different ways. Use play money to help.

____

Solve.

Roast Beef Sandwich	$6.79
Grilled Cheese Sandwich	$3.49
Salad	$3.79

5. Leroy bought a roast beef sandwich for lunch. He paid the cashier and received $13.21 in change. How much did Leroy give the cashier to pay for his lunch?

____

6. Pablo paid for his lunch with a $5 bill. He received a $1 bill, 2 quarters, and 1 penny in change. How much change did he receive? __________

What did Pablo have for lunch?

____

Count Money and Make Change

Solve.

1. Mr. Smith sold a $0.50 fruit bar to Molly. She gave him a $1 bill. How much change should Molly get?

____

2. Suppose you buy a book and get $1.57 in change. What bills and coins could the change be?

____

Solve.

3. Amelia bought a $4.36 magazine and paid with $10 bill. How much change does she get back? List the bills and coins of the change.

____

4. Maria paid for a board game with a $10 bill. She got $1.68 in change. How much did the board game cost?

____

Solve.

5. Antonio pays for his textbook with a $10 bill. He gets back a $1 bill, 1 quarter, and 2 dimes in change.

   How much change did he get?

   ____

   How much did the textbook cost?

   ____

6. Josh sold a CD to Vera for $5.55. Vera gave Josh a $10 bill for the CD. Josh has no quarters or dimes, but gives Vera the correct amount of change. Tell what bills and coins he may have given her.

____

Problem Solving: Skill
Using the Four-Step Process

Solve.

1. Jen lives 400 miles from Grandpa. Randy lives 100 miles more from Grandpa than Jen. How many miles does Randy live from Grandpa?

____

2. A water toy costs $2.75. How much change will you get if you pay for the toy with a $5 bill?

____

Solve.

3. Students can earn 200 points toward a prize for each magazine subscription they sell. If a student sells 3 subscriptions, how many points does he or she get?

____

4. Ms. Ruiz paid for her groceries with a $10 bill and a $5 bill. She got $2.78 change. How much did the groceries cost?

____

Solve.

5. Ed gets a raise of $1,000 each year. When Ed started working at Foster's Flowers, he earned $15,000. This year, he will make $19,000. How many years has Ed been working at Foster's?

____

6. Maggie paid for a book with 2 bills. Both bills had the same value. The cashier gave Maggie $2.48 in change. The book cost less than $12. What 2 bills did Maggie pay with?

____

How much did the book cost?

____

2.1 Compare Numbers and Money

Solve.

1. Eli has read 110 pages of his book for his book report. Dee has read 101 pages. Who has read more pages so far?

   ____

2. In the school bookstore, a pen costs $2.34. A large notebook costs $2.40. Which costs less, the pen or the notebook?

   ____

Solve.

3. Mrs. Jones gave her class a multiple-choice test. There were 175 questions on the test. Anna answered 168 questions correctly. Jorge answered 159 questions correctly. Who scored higher on the test? ____________

   How can you tell?

   ____

4. The car Manuel wants to buy costs eight thousand, five hundred nine dollars at Corey's Car Dealership. It costs $8,590 at Ned's Cars and Trucks. At which car dealership should Manuel buy his car? Tell why.

   ____

   ____

Solve.

5. Paul has $5.26, Kim has $6.25, and Jara has $5.52. They each want to buy a salad for lunch. The salad costs $5.30. Who can buy the salad? ____________

   How can you tell?

   ____

   ____

   ____

6. Music Town sold 1,252 CD players in January. They sold one thousand, two hundred ninety-eight in February. In March, they sold 1,521 CD players. They sold fewer than 1,000 in April. In which month did they sell the most? ____________

   In which month did they sell the fewest?

   ____

Order Numbers and Money

Solve.

1. Grade 2 has 134 students. Grade 3 has 143 students. Grade 4 has 129 students. Write the grades in order from greatest to least number of students.

____

2. A pound of potatoes costs $2.19. A pound of onions costs $0.99. A pound of tomatoes costs $2.29. Write the foods in order from least to most expensive.

____

Solve.

Car Type	Cost	Number Sold
Sports Car	$9,708	2,190
Sedan	$9,679	1,987
Compact	$8,998	2,910

3. According to the table, which car costs the most money?

____

4. Write the names of the cars in order from the least number of cars sold to the greatest number of cars sold.

____

Solve.

5. Gordon has $8.78 in his wallet. Beth has $8.87 in her wallet. Craig has more than Gordon, but less than Beth in his wallet. How much money could Craig have in his wallet?

____

6. The population of four towns are as follows: 3,154; 3,451; 3,514; 3,415. Newtown has more people than Oldtown. Middletown has the largest population and Youngstown has more people than Newtown. Match the towns to their populations.

____

Estimate Quantities

Solve.

1. Justin has a small bookshelf over his bed with his favorite books on it. Would the shelf have about 10 books, 100 books, or 1,000 books?

____

2. Use Bag A as the benchmark number. Are there about 20 grapes or 80 grapes in Bag B?

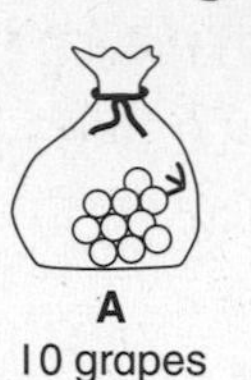

A
10 grapes

B

____

Solve.

3. Jason filled his pants pockets with as many nickels as they could hold. Could Jason's pocket have 20 nickels, 200 nickels, or 2,000 nickels?

____

4. Use Box A as the benchmark number. Are there about 60 pens or 200 pens in Box B?

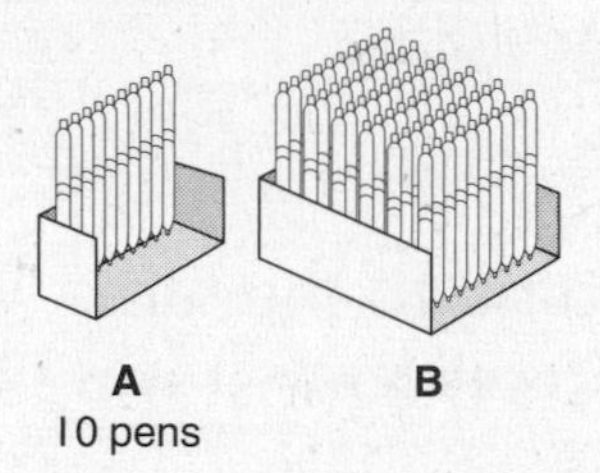

A
10 pens

B

____

Solve.

5. A stack of 20 pennies is a little taller than 1 inch. Brandon has a stack of pennies that is a little more than 3 inches high. About how much money does Brandon have?

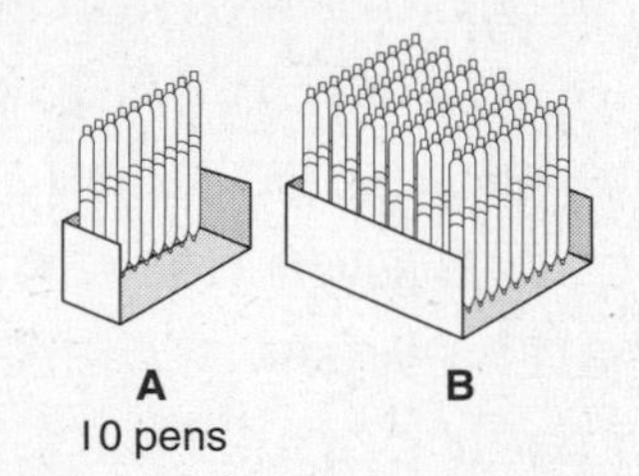

A
10 pens

B

____

6. The 20 students in Mr. Fargo's 3rd-grade class each have a box of crayons. If they put all their crayons in a pile on a table, will there be about 20 crayons, 200 crayons, or 2,000 crayons on the table?

____

Round to Tens and Hundreds

Solve.

1. It takes Alan exactly 11 minutes to walk to school each day. About how long does it take to the nearest ten?

____

2. Yumi's dog, Max, weighs 59 pounds. About how much does Max weigh to the nearest ten?

____

Solve.

3. A 36-inch T.V. costs $509 at Sam's Electronics. This month it is on sale for $449. To the nearest hundred dollars, how much is the television on sale? __________

   How much was it, to the nearest hundred dollars, before the sale?

____

4. The Sears Tower in Chicago is 1,450 feet tall. How tall is the tower to the nearest hundred feet?

____

Solve.

5. Kyle has $1,861 in his bank account. His sister, Kayla, has about $400 less in hers. How much does Kayla have in her account to the nearest hundred dollars?

____

6. Which 3-digit numbers round to 500 when rounded to the nearest hundred, and also round to 460 when rounded to the nearest ten?

____

Round to the Nearest Thousand

Solve.

1. The Hansons' new refrigerator cost $1,085. How much did the refrigerator cost to the nearest thousand dollars?

____

2. It is 1,845 miles between Los Angeles and St. Louis. What is the distance between the cities to the nearest thousand miles?

____

Solve.

3. The Morrisons paid $10,825 to have their kitchen remodeled. The Wongs paid $400 less. How much to the nearest thousand dollars did the Wongs pay to have their kitchen remodeled?

____

4. The largest city Hector has ever visited had a population of 79,702 people. What is this city's population to the nearest thousand?

____

Solve.

5. The top-selling car in the year 2000 had sales of 422,961 cars. This was 18,000 more cars sold than the second top-selling car that year. To the nearest thousand, about how many second top-selling cars were sold in 2,000?

____

6. Mr. and Mrs. Garcia bought a new house. If the exact price of the house is rounded to the nearest ten, the Garcias paid about $223,500. If the exact price is rounded to the nearest hundred, the price is also about $223,500.

   What are the possible **exact** prices of the new house?

____

Problem Solving: Strategy Make a Table

Use the data and make a table. Solve.

1. How many members are there in all in the Chess Club and the Math Club?

____

2. How many more members does the Math Club have than the Rocket Club?

____

After-School Club Members	
Chess Club:	Pablo, Sam, Gary, Justin,
Rocket Club:	Brittany, Maddy, Joe, Greg
Math Club:	Jara, Parth, Will, Howie, Gina, Lily

Use the data to make a table. Solve.

3. How many more students are in the Cougars than the Jaguars?

____

4. The teacher moved Keisha and Dean into the Jaguars. Now how many students are in the Cougars?

____

Reading Groups	
Tigers:	Ellen, Beth, Mary, Pete, Juan, Sara, Jose, Larry, Ty
Jaguars:	Fran, Mara, Ezra, Lee, Paulo
Cougars:	Dean, Darryl, Cara, Keisha, Mei, Bo, Ed, Liz, Pat, Will, Pedro

Use the data to make a table. Solve.

5. If each student scores 3 points higher on the next math test, which students will score in the 90s?

____

6. If you round each score to the nearest ten, how many more students scored in the 90's than in the 80's?

____

Here are the top grades on the last math test in Mr. Moreno's class:
Seth: 85; James: 97; Neil: 96; Jan: 82; Mike: 78; Gail: 88; Josh: 90; Randy: 85

Algebra: Addition Properties

Solve.

1. Lin works 3 hours after school on Monday. She works 3 hours on Wednesday and 4 hours on Friday. How many hours did Lin work this week? ______

   Which Addition Property did you use to solve?

   ______

2. There are 8 pennies and 9 dimes on the table. How many coins in all are on the table? ______

   Which addition strategy did you use to solve?

   ______

Solve.

3. Rod ran a total of 8 miles last week. He will run 2 more miles this week than last. How many miles will he run in all during both weeks?

   ______

4. One page in a sticker album has 3 rainbows, 7 hearts, 4 flowers, and 4 stars. How many stickers in all are on the page?

   ______

Solve.

5. Lee scored 4 points in Monday's game, and 5 points in Tuesday's game. He hopes to score 2 more points in Friday's game than he did on Monday. If he does, how many total points will he have scored for the 3 games?

   ______

6. The school Bake Sale lasted 3 days. On the first day, Margo sold 4 cupcakes. On day two, she sold 6 cupcakes. On day three, she sold as many cupcakes as she had sold on days one and two together. How many cupcakes did she sell in all?

   ______

Algebra: Addition Patterns

3.2

Solve.

1. There are about 400 boys and 300 girls at the movie theater. About how many children are there in all?

   ____

2. A box of beads has 20 red beads, 40 blue beads, and 60 white beads. How many beads are in the box?

   ____

Solve.

3. Stan and Jenna were working on a puzzle. Stan fit 308 pieces into the puzzle. Jenna fit the remaining 50 pieces to finish the puzzle. How many pieces are in the puzzle?

   ____

4. The Smiths drove 300 miles on the first day of their trip. They drove 100 more miles on the second day than on the first. On the last day of the trip, they drove another 300 miles. How many miles did they drive in all?

   ____

Solve.

5. A soap company delivered 1,200 bars of soap to a grocery store. The store sold 600 bars in January, 400 in February, and the rest of the soap in March. How many bars of soap were sold in March?

   ____

6. The members of the Science Club collect bottles to recycle. In September, they collected 2,000 bottles. In October, they collected 1,000 more than in September. In each of the next two months, they collected 1,000 more bottles than the month before. How many bottles in all did the club collect over the four months?

   ____

3.3 Explore Regrouping in Addition

Solve. Use place-value models.

1. There are 18 crayons in one box and 24 crayons in another box.

tens | ones

How many crayons in all? ______
Did you regroup the ones? ______
Did you regroup the tens? ______

2. Harry has 161 baseball cards. He buys a pack of 42 cards.

hundreds | tens | ones

How many cards does Harry have now? ______
Did you regroup the ones? ______
Did you regroup the tens? ______
Did you regroup the hundreds? ______

Solve. Use place-value models if you need help.

3. If there are 365 days in a year, how many days are in two years?

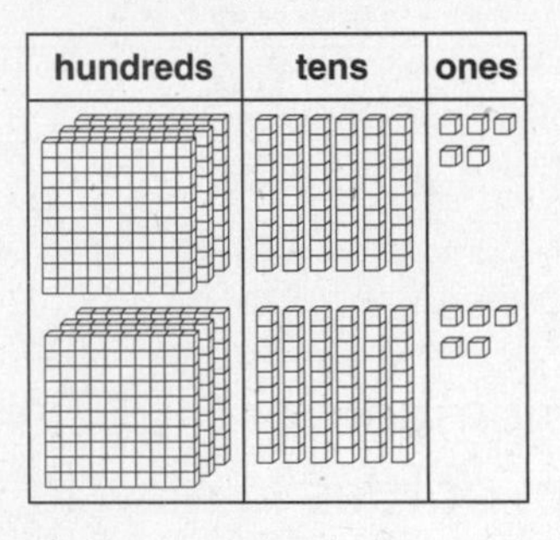

______

4. A concert hall has 624 seats on the first level and 509 seats on the second level. How many seats are there in all in the concert hall?

______

Solve.

5. On Monday, 748 people rode on the merry-go-round. On Tuesday, 965 rode on it. How many riders were there in all for both days?

______

6. Play Park had 650 visitors. On the next day there were 158 more visitors than the day before. How many visitors were there in all for both days?

______

Add Whole Numbers

Solve.

1. For his lunch, Jack paid $2.09 for a hamburger and $1.75 for fries. How much did Jack's lunch cost?

____

2. Indi made one phone call and talked for 68 minutes. She made another call and talked for 54 minutes. How many minutes in all did she talk on the phone?

____

Solve.

3. Super Computer Store is selling a laptop for $729. Laptop World is selling the same laptop for $92 more. How much does the laptop computer cost at Laptop World?

____

4. A mystery book costs $9.55. A dictionary costs $8.75. How much will you pay if you buy both books?

____

Solve.

5. Ira has $5.85. Sara has $1.25 more than Ira. They want to buy a scarf for Tom that costs $15.00. If the two children combine their savings, can they buy the scarf? Explain why or why not.

____

6. There are 1,250 coins in a drawer at the bank. There are quarters, dimes, and nickels. The bank teller knows that there are 869 dimes and 370 nickels. How many quarters are in the drawer?

____

3.5 Estimate Sums

Estimate.

1. There are 43 cats and 48 dogs in the animal shelter. Are there more than 100 animals or fewer than 100 animals in the shelter?

____

2. A pet store has 192 fish in a large tank in the front of the store. Another tank in the back of the store has 213 fish. About how many fish are there in all?

____

Estimate.

3. Terry travels 72 minutes to and from school each day. About how many minutes in all does Terry travel each day?

____

4. The three salespeople in a car dealership sold the following number of cars last year: 112 cars, 98 cars, and 109 cars. Did they sell more or fewer than 200 cars in all? Explain how you can estimate to find out.

____

Estimate.

5. Cedar School has about 1,500 students. Pine Grove has about 1,600 students. There are 712 girls in one school and 789 girls in the other. There are 687 boys in one school and 878 boys in the other. Match the numbers of boys and girls to their correct school.

____

6. Keisha had 721 pennies in her bank. She saved 352 more pennies this month. Jared has a total of 1,263 pennies. Who has more money? Explain how you used estimation.

____

Problem Solving: Skill Estimate or Exact Answer

Use the data to solve.

1. Are there more or fewer than 400 kites? Explain why your answer is an estimate or exact answer.

____

2. Mr. Hollis bought all of the black and green kites. How many kites in all did he buy?

____

Number of Kites

Red	92
Blue	87
Black	79
Green	98

Use the data to solve.

3. About how many more sports and mystery books are there than books about art and music?

____

4. The librarian wants to move all the art and music books. How many books in all will she move?

______________ books

Books in the Library

Sports	248
Art	188
Mysteries	259
Music	63

Use the data to solve.

5. About how much greater is the area of Fresno County than the areas of Amador and Marin added together? about __________ sq mi. Explain.

____

6. Solano County is 236 square miles greater than the county in which Tim lives. Where does Tim live?

____

Areas of California Counties

Alameda	738 sq mi.
Fresno	5,963 sq mi.
Monterey	3,322 sq mi.
Amador	593 sq mi.
Marin	520 sq mi.
Solano	829 sq mi.

Add Greater Numbers

 **Solve.**

1. The school music department bought a new piano and a new flute. The piano cost $1,520, and the flute cost $650. How much did the music department spend in all?

____

2. Music students sold 1,325 tickets for the Spring Concert and 1,625 tickets for the Fall Concert. How many tickets were sold in all?

____

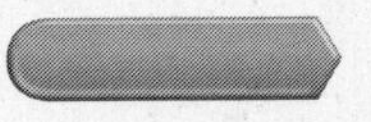 **Solve.**

3. Last month Hugo earned $1,089. This month he will earn $488 more than last. How much will Hugo earn for the two months of work?

____

4. A video game costs $39.99 and a game pad costs $29.99. Nate does not want to spend more than $69. Can he buy both items? Explain why or why not.

____

____

 **Solve.**

5. The Hospital Toy Fund collected $578.85 last week. If it collects $25.50 more this week than last, can the fund buy more than $1,100 worth of toys? Explain.

____

____

6. A local company donated $247,500 to the hospital last year. Now the hospital needs only $85,000 more for a new X-ray machine. How much does the new machine cost?

____

Add More Than Two Numbers

Solve.

1. A bowl of fruit punch contains 124 ounces of grape juice, 64 ounces of apple juice, and 108 ounces of pear juice. How many ounces of juice are in the bowl?

____

2. The cook ordered fresh fruit for school lunch on Monday. She ordered 119 apples, 78 bananas, and 98 oranges. How many pieces of fruit did she order for lunch on Monday?

____

Solve.

3. The office manager puts 3 boxes of pens into the supply closet. Each box has 144 pens. How many pens are in the supply closet?

____

4. A secretary gets 3 new boxes of staples. Each box has 1,250 staples. She still has 450 staples left in her old box. How many staples does she have in all?

____

Solve.

5. A small box of craft sticks has 250 sticks. A medium box has 150 more than small. A large box has 1,100 more than medium. How many craft sticks are there in the small, medium, and large boxes added together?

____

6. Grades 1, 2, 3, and 4 made a large collage in the hall. The collage has 1,000 photos of school activities. Grade 1 put 77 photos in the collage. Grade 2 put in 358 photos and Grade 4 put in 288 photos. How many photos of Grade 3 activities are in the collage?

____

4.3 Problem Solving: Strategy
Draw a Diagram

Draw a Venn diagram to solve.

1. There are 8 students in a math group using pattern blocks. Three students have squares. Two have triangles. Three have both squares and triangles. How many have triangles?

____

2. There are 10 students in a reading group. Five students have red bookmarks. Two have white bookmarks. Three have both red and white bookmarks. How many students have red bookmarks?

____

Draw a Venn diagram to solve.

3. All 25 students in Meena's class own video games. Ten own sports games. Six own racing games. Nine own both sports and racing games. How many students own sports games? How many own racing games?

____

4. There are 120 third grade students at Parks School. Forty of them own a computer. Fifty-three of them own a calculator. Twenty-seven of them own both. How many students own computers? How many own calculators?

____

Draw a Venn diagram to solve.

5. A store shelf has 15 mugs with smiley faces on them. There are 18 mugs with rainbows. The rest of the mugs have both smiley faces and rainbows on them. If there are 40 mugs on the shelf, how many have both smiley faces and rainbows on them? How many mugs have smiley faces?

____

6. Misha asked her classmates if they have a brother, a sister, or both. There are 27 students in her class. Nine students have only a brother. Misha and 6 other students have both a brother and a sister. How many students have only a sister?

____

Choose a Computation Method

Solve. Use paper and pencil, a calculator, or mental math.

1. Pam rides on the bus for 18 minutes, then walks 12 more minutes to get to school. How long does it take Pam to get to school? Which method did you use to solve?

____

2. At a garage sale Emily buys a table for $7.75 and a chair for $4.18. How much did Emily spend in all? Which method did you use to solve?

____

Solve. Use paper and pencil, a calculator, or mental math.

3. The art teacher has 6 red, 8 blue, and 4 white jars of paint on the top shelf of her cart. She has 12 more jars of paint on the bottom shelf. How many jars of paint does she have in all? Which method did you use?

____

4. Super Sal's Auto Shop sold 1,568 tires in each of the last two years. This year 56 more tires were sold than last year. How many tires in all have been sold in the last three years? Which method did you use?

____

Solve. Use paper and pencil, a calculator, or mental math.

5. Mr. Roe paid $2,683 for his computer. Mr. Baglio paid about $500 less for his. How much did Mr. Baglio pay for his computer to the nearest hundred dollars?

____

6. There are 1,195 seats in the auditorium. At a play, there were 435 women and 687 men in the seats. The rest of the seats were empty. How many were empty?

____

5.1

Algebra: Relate Addition and Subtraction

Solve.

1. There are 14 children in line to buy tickets to a movie. Six of the children are boys. How many girls are in line? _____ girls
Write a related subtraction fact you can use to find □ in 6 + □ = 14.

____

2. Sam is carrying 4 grocery bags. His little brother Carl is carrying the rest. The boys are carrying 7 bags in all. How many bags is Carl carrying? _____ bags
Write a related addition fact you can use to find □ in 7 − □ = 4.

____

Solve.

3. There are 17 dog biscuits left in the box. Some are thick and others are thin. Eight of the biscuits are thick. What fact family can you use to find out how many are thin?

____

____

How many are thin? _____

4. If 5 of the 12 members on the basketball team have new sneakers, how many do not have new sneakers? _____
Write the related facts you used to find out.

____

____

Solve.

5. Tanya added 6 new photos to her photo album page. Now the page is full with 18 photos. How many photos were on the page before Tanya added 6 new photos? _____ photos

6. On a math test, students were asked to name 3 numbers in the same fact family. Lisa answered 7, 5, and 10. Jeremy answered 7, 4, and 11. Which student's answer is correct? Explain.

____

____

Problem Solving: Skill
Identify Extra Information

Circle the extra information. Then solve.

1. Danny is reading a 230-page book. He read 52 pages last night. He read another 39 pages today. How many pages has Danny read so far?

   ________ pages

2. Hanna and Grace each bought one issue of *Wild Animals*. The magazine costs $5 each issue and $34.95 for the year. How much did Hanna and Grace pay all together?

   ____

Identify the extra information. Then solve.

3. A fish tank has the same number of guppies and mollies in it. Guppies cost $1.25 each and mollies cost $3.25 each. If there are 14 fish in all in the tank, how many of each type are there?

   ________ of each type

4. Pedro bought a new fish tank for $9.99 and a filter for $5.95. The tank holds 10 gallons of water. If he paid $0.95 tax for the items, how much did Pedro spend in all? $__________

Identify the extra information. Then solve.

5. Last month, the bank paid Shandi $12.75 in interest on her bank account. This month, she earned $2.64 more in interest than last month. Shandi has $9,000 in her bank account. How much interest did she earn in the last two months? $__________

6. Mr. Burns bought 20 shares of XYZ stock for $1,558. He paid a $25 fee to his stock broker. Mrs. Rogers bought the same stock, but paid only $15 to her stock broker. What was the total amount that Mrs. Rogers paid to buy the stock? $__________

5.3 Algebra: Subtraction Patterns

Solve.

1. A large above-ground pool holds 900 gallons of water. A smaller one holds 500 gallons. How many more gallons of water does the large pool hold than the smaller one?

_________ gallons more

2. Antonio played in the pool for 60 minutes. His younger sister, Julia, played in the pool for 20 minutes. How much less time was Julia in the pool than Antonio?

_________ minutes less

Solve.

3. A carpenter bought a box of nails. The box contains 1,000 nails. Some of the nails are 2 inches long, and the rest are 1 inch long. If there are 300 two-inch nails, how many one-inch nails are in the box?

_________ one-inch nails

4. One building is 1,100 feet tall. The building next to it is 800 feet tall. Is the taller building more or less than 200 feet higher than the shorter building?

_________ than 200 feet higher

Explain. ______________________

Solve.

5. Mrs. Lopez had $500 to spend on Dario's birthday presents and party. She bought him a bike that cost $244, and a helmet that cost $56. How much does Mrs. Lopez have left to spend on the party?

$_____________ left

6. There are 3,000 people in the audience.
There are 1,250 men, 965 women, and 485 boys.
How many girls are in the audience?

_____________ girls

Explore Regrouping in Subtraction

Use the place-value models to solve.

1. There were 175 peaches at the fruit stand.

   Customers bought 82 of the peaches. How many peaches are left? ________ peaches

   Did you need to regroup ones? ____ tens? ____

2. Another crate has 272 red and green apples.

   There are 123 red apples in the crate. How many apples are green? ________ green apples

   Did you need to regroup ones? ____ tens? ____

Use the place-value models to solve.

3. Tanisha bought a pack of 225 sheets of paper for her homework.

   After a week, she has 198 sheets of paper left. How many sheets of paper did Tanisha use?

   ________ sheets

4. The school library would like to raise $915 to buy more books.

   So far, the library has raised $475. How much more money does the library need to reach its goal?
   ________ more

Use place-value models to solve.

5. The health food store had 254 granola bars. They sold 85 bars yesterday and another 78 bars today. How many granola bars does the store have left?

   ________ granola bars

6. Evan has 85 baseball cards and 129 basketball cards. Alan has 312 football cards. Who has more cards in all? ____

   How many more cards?
   ________ more cards

Subtract Whole Numbers

Use the data in the table to solve the problems:

School Cafeteria Lunch Menu	
Food Item	Price
Slice of Pizza	$1.69
Bowl of Chicken Soup	$1.45
Grilled Cheese Sandwich	$2.25
Chicken Fingers	$2.75
Fries	$2.15
Salad	$1.75
All Drinks:	$0.79

Solve.

1. How much more does a grilled cheese sandwich cost than a bowl of chicken soup?
$________

2. How much money will you save if you buy a salad instead of fries?
$________

Use the menu above to solve.

3. Roy has $9.90. He buys chicken fingers and a glass of milk. How much does Roy have left?
$________

4. How much more does an order of fries and a drink cost than just a salad?
$________ more

Use the menu above to solve.

5. Chris bought a salad and a glass of juice. Tara bought a slice of pizza, soup, and a glass of milk. Who spent more money on lunch? ________
How much more? $________

6. Aidan gave the cashier $3.00 for a lunch item and a drink. He got back 76¢ in change. What did Aidan buy?
____

Regroup Across Zeros

Solve.

1. The best bowler in the Junior Bowler's League scored 150 points. Jason scored 125 points. How many points higher did the best bowler score than Jason?

   _____ points higher

2. There are 70 bowlers in the league this year. There were only 54 bowlers last year. How many more bowlers joined the league this year?

   _____ more bowlers

Votes for School President	
Candidate	Number of Votes
Ariana	200
Miguel	147
Tyrone	171

Use the chart to solve.

3. How many more votes did the winner get than Miguel?

   ________ more votes

4. How many more votes did Tyrone need to win the election?

   ________ more votes

Solve.

5. Harrison and Jordan played 3 computer games. Jordan scored 124 points in the first game and 268 points in the second game. Harrison scored a total of 600 points for all 3 games. How many points does Jordan need in the third game to beat Harrison's score? ________ points

6. Keisha is saving money for a new computer that costs $480. She has saved $175. She found a coupon for $50 off the price of the computer. How much more money does Keisha need to save to buy the computer?

   $________ more

Estimate Differences

Use estimation to solve.

1. A basketball coach won 132 games. He won 79 more games than he lost. About how many games did he lose?

   about ________ games

2. The bleachers in the gym can seat about 289 people. If there are 191 people in the gym, about how many more people can fit in the bleachers?

   about ________ more

Use estimation to solve.

3. The Chase Tower in Houston, Texas, is 1,002 feet tall. It is 499 feet taller than the Huntington Building. About how tall is the Huntington Building?

   about ________ feet tall

4. The Bank One Center in Dallas is 787 feet tall. It is 304 feet taller than Harwood Center. Is the Harwood Center greater than or less than 500 feet tall?

   ________ than 500 feet tall
   Explain your answer.

   ____

   ____

Use estimation to solve.

5. Mr. Frasier has $875 in his savings account and $989 in his checking account. Mr. Willis has $1,289 in his savings account and $497 in his checking account. Who has more money in all?

   ____

   About how much more?
   about $________ more

6. Arco Company has $8,050 to spend on office equipment. The company bought a computer for $2,885, a copy machine for $485, and a video camera for $1,005. If the company buys another computer, about how much will be left in the budget?

   about $________

Problem Solving: Strategy Write an Equation

Write a number sentence to solve.

1. Robert is 47 inches tall. His older brother Randy is 65 inches tall. How much taller is Randy than Robert? ________ inches taller
Write an equation.

____

2. Robert weighs 52 pounds. Randy weighs 68 pounds more than Robert. How much does Randy weigh? ________ pounds
Write an equation.

____

Write a number sentence to solve.

3. A new house has 1,252 square feet of floor space. The owner plans to cover 590 square feet with carpeting and the rest with hardwood. How many square feet of hardwood will the owner need? ________ square feet
Write an equation.

____

4. A can of paint can cover 475 square feet. If the painter has 2 cans of paint, can he cover 900 square feet of walls? ________
Write an equation. Explain your answer.

____

____

Write a number sentence to solve.

5. Heather and Aaron each bought a set of markers. Heather paid $3.15 for her set. Aaron paid with a $5 bill. He got back $2.35 in change. How much more were Heather's markers than Aaron's?
$________ more
Show equations.

____

____

6. There are 500 sheets of art paper in a pack. The pack has 125 white sheets, 135 black sheets, and 115 yellow sheets. The rest of the sheets are red. How many red sheets of art paper are in the pack? ________ red sheets
Show equations.

____

____

6.4 Subtract Greater Numbers

Solve.

1. A library has 2,220 books about sports and 1,814 books about animals. How many more sports books are there than animal books?

   ________ more books

2. At the book sale, Sharika paid for a book with a $5-bill. She got back $3.85 in change. What was the price of the book?

   $________

Solve.

3. In the 2000 NFL season, Eddie George rushed for 1,509 yards. Lamar Smith rushed for 1,139 yards, and Corey Dillon rushed for 1,435 yards. How many more yards were rushed by Eddie George than Corey Dillon?

   ________ more yards

4. Pittsburgh University won the college football championship in 1937. They won again in 1976. How many years were there between championships?

   ________ years

Solve.

5. Jeff bought two tickets to a baseball game. Each ticket cost $12.75. He paid with $30. How much change should he get back?

   $________

6. The baseball stadium has 48,500 seats. One week before the game there were still 9,956 seats left. The day of the game there were 4,089 seats left. How many seats were sold during the week before the game?

   ________ seats

   How many seats were sold by the day of the game?

   ________ seats

Choose a Computation Method

Use paper and pencil, a calculator, or mental math to solve.

1. Leon has a collection of 150 music CDs. He gave 27 to his friends. How many CDs does he still have?

   _______ CDs _______ method

2. Mandy spent $1.50 in one store and $1.40 in another. How much did she spend in all?

   $_______ _______ method

Use paper and pencil, a calculator, or mental math to solve.

3. On Monday 3,416 cars traveled on Wellcott Avenue. On Tuesday 2,987 cars traveled on Wellcott Avenue. How many cars traveled on the road during the two days?

   _______ cars _______ method

4. On Wednesday and Thursday a total of 8,450 cars traveled on the freeway. If 5,100 traveled on the freeway on Wednesday, how many cars traveled the road on Thursday?

   _______ cars _______ method

Use paper and pencil, a calculator, or mental math to solve.

5. Bianca's mom took her and a friend to the roller rink. Tickets cost $12.50 an hour for adults and $8.95 an hour for children. She will pay for 1 adult and 2 children for 1 hour. She gave the cashier $50. How much change will Bianca's mom get back?

   $_______ _______ method

6. Three teams in the rocket club launched rockets. The Blue Team's rocket went 967 feet in the air. The Red Team's went 833 feet high. The Green Team's went 800 feet high. How much higher must the Green Team's rocket go if its goal is to go 33 feet higher than the Blue Team's?

   _____ feet higher _____ method

Tell Time

A

B

C

D

E

F

Use the clocks to solve. Write A.M. or P.M.

1. Clock A shows the time school starts each morning. What time does school start? ____________

2. Clock B shows when school lunch ends. What time does lunch end in school? ____________

Use the clocks above to solve. Write A.M. or P.M.

3. Clock C shows the time the sun will rise tomorrow morning. Write the time the sun will rise.

   ____

   Then show a different way to write the time.

   ____

4. Clock D shows when Seth takes his daytime nap. Write the time he takes his nap.

   ____

   Then show a different way to write the time.

   ____

Use the clocks above to solve. Write A.M. or P.M.

5. Clock E shows when Gerry starts preparing supper. Write the time the clock shows. ____________

   Then show a different way to write the time.

   ____

6. Clock F shows the time Mr. Willis went to bed after watching television. Write the time the clock shows. ____________

   Then show a different way to write the time.

   ____

Convert Time

7.2

Solve.

1. James spent $\frac{1}{2}$ hour on his math homework and 15 minutes on spelling. For how many minutes in all did James do homework?

   _____ minutes

2. The basketball game lasted for 2 hours. How many minutes did the game last?

   _____ minutes

Solve.

3. Rolanda visited her grandmother. She stayed for an hour and ten minutes. How many minutes in all did the visit last?

   _____ minutes

4. It took a painter an hour to paint the living room and 45 minutes to paint the kitchen. How much longer did it take to paint the living room than the kitchen? Tell the time in minutes and in hours.

   _____ minutes or _____ hour

Solve.

5. Tom works for 3 hours every day after school. Keesha works for 210 minutes after school. Who works longer?

   ____

   How much longer? Tell the time in minutes and in hours.

   _____ minutes or _____ hour

6. Jesse swam the race in 124 seconds. Her teammate Carrie swam the race in 2 minutes and 10 seconds. Who won the race?

   ____

   By how many seconds?

   by _____ seconds

Elapsed Time

Solve.

1. Every day, Louis practices the piano from 3:15 P.M. to 4:00 P.M. How long does Louis practice each day?

____

2. On Saturday afternoon, Louis checked the time. His watch read:

His piano lesson begins in 3 hours. What time will his lesson begin?

____

Solve.

3. The after-school Math Club meets from 3:15 P.M. to 6:00 P.M. How long does the meeting last in hours and minutes?
______ hours and ______ minutes
How many minutes in all does it last? ______ minutes

4. David started his homework at 4:10 P.M. He worked for 1 hour and 15 minutes. Then he ate supper. At what time did David eat supper?

____

Solve.

5. School starts at 8:30 A.M. Science class starts 15 minutes later. If science class ends at 9:35 A.M., how long is science class?

____

6. The music class starts at 2:30 P.M. Music lasts for 45 minutes. Ten minutes later is the end of the school day. Margo eats dinner 2 hours and 20 minutes after school ends. At what time does Margo eat dinner?

____

Calendar

October							November						
S	M	T	W	T	F	S	S	M	T	W	T	F	S
			1	2	3	4							1
5	6	7	8	9	10	11	2	3	Election Day 4	5	6	7	8
12	Columbus Day 13	14	15	16	17	18	9	10	Veteran's Day 11	12	13	14	15
19	20	21	22	23	24	25	16	17	18	19	20	21	22
26	27	28	29	30	31		23	24	25	26	Thanks-giving 27	28	29
							30	31					

Use the calendars to solve.

1. If today is October 23rd, how many weeks is it until Thanksgiving?

   ______ weeks

2. The Garden Club meets on the second Tuesday of each month. On what date will they meet in October?

   ____

Use the calendars above to solve.

3. How many weeks and days are there from Election Day to Thanksgiving?

   ______ weeks and ______ days

   How many days in all are there?

   ______ days

4. Lana starts food shopping for Thanksgiving 3 weeks before the Friday after Thanksgiving. On what date does Lana start her food shopping?

   ____

Use the calendars above to solve.

5. Cheryl's birthday is 3 weeks and 3 days after Columbus Day. What date is Cheryl's birthday?

   ____

   On what day does it fall this year?

   ____

6. Emilia will run in a marathon on the third Sunday in November. She will need to start training for the race 42 days before. On what day and date should Emilia start training?

   ____

7.5 Time Lines

Important Events in Casey's Life

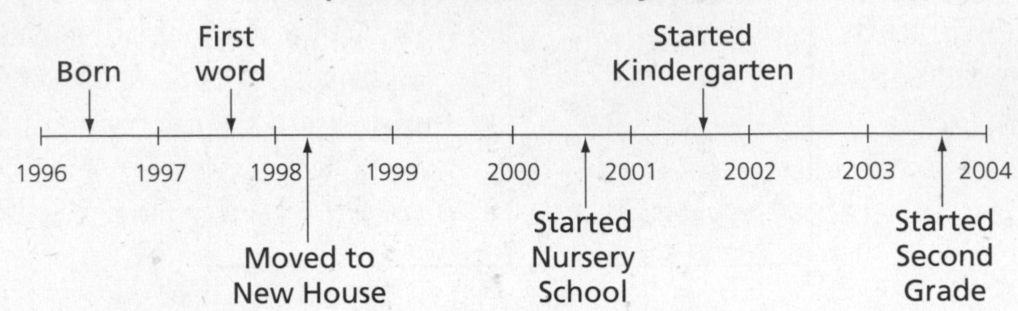

Use the time line about Casey's life to solve.

1. In what year was Casey born?

____

2. How many years after she was born did Casey move to a new house?

____

Use the time line above to solve.

3. Casey's baby brother was born the year after she moved. In what year was Casey's brother born?

____

4. What happened between 2001 and 2002?

____

Use the time line above to solve.

5. Casey's sister Linda is 3 years older than Casey. In what year was Linda born?

____

6. In what year would you predict Casey started third grade?

____

Problem Solving: Skill
Identify Missing Information

Solve or identify missing information.

1. Dr. Ortiz starts work in the hospital at 8:00 A.M. After her work in the hospital, she goes to her office. How long does Dr. Ortiz work at the hospital?

____

2. Mr. Miller starts work at 9:30 A.M. He has a doctor's appointment 3 and one-half hours after he starts work. What time is Mr. Miller's doctor's appointment?

____

Solve or identify missing information.

3. Maddie went to band practice from 3:30 P.M. to 4:45 P.M. and then went home. It took Maddie 25 minutes to get home. What time did she get home?

____

4. Maddie starts dinner at 5:30 P.M. She plans on watching her favorite TV show at 8:00 P.M. How long must Maddie wait after dinner to watch the show?

____

Solve or identify missing information.

5. On Mondays, Anna works from 8:15 A.M. to 4:30 P.M. On Fridays, she starts at 9:15 A.M. and works 2 hours less than on Mondays. When does Anna finish work on Fridays?

____

6. Eric wakes up 2 hours and 15 minutes before he leaves for work. It takes him 1 hour to shower and dress, and 1 hour and 15 minutes to make and eat breakfast. He starts work at 10:00 A.M. and finishes at 5:45 P.M. What time does Eric wake up ?

____

Collect and Organize Data

Make a chart or line plot. Use a separate sheet of paper.

1. Make a tally chart to show the following data:
Weight in pounds of Pot Belly Pigs:
10, 15, 15, 20, 20, 20, 25, 25, 25, 15, 15, 30, 20, 30, 20

____

2. Make a line plot to show the data in the tally chart you made.

____

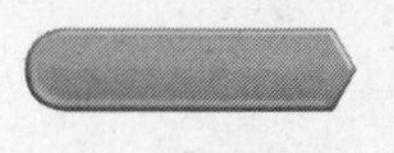

Make a tally chart or line plot. Use a separate sheet of paper.

3. Ages of the runners in the race:
14, 14, 26, 21,15, 23, 17, 17, 18, 23, 18, 18, 26, 25, 14, 16, 16, 19, 21, 21, 25, 19, 19, 17, 15, 15, 25, 23, 26, 14
The greatest number of runners are what age? _____
The least number of runners are what age? _____

4. How many more runners are 14 to 19 years old than 20 to 26 years old?

_____ more

Make a tally chart or line plot. Use a separate sheet of paper.

5. Heights in inches of 3rd-grade students: 48, 46, 52, 51, 47, 46, 48, 52, 57, 45, 48, 49, 50, 46, 50 , 52, 47, 50, 52, 48, 49, 49, 50, 52, 50.
How many students are 3 to 5 inches taller than the shortest student? _____

6. Suppose all of the students grow 2 inches next year. How many more students will be 50 inches or taller than this year?

_____ more

Find Median, Mode, and Range

8.2

Solve.

1. Five students read 4, 2, 3, 2, and 6 books last month. What is the mode, median, and range for the data?

   Mode: ____ Median: ____
   Range: ____

2. The soccer team made these field goals in 7 games: 1, 4, 4, 5, 3, 4, 3. What is the mode, median, and range for the data?

   Mode: ____ Median: ____
   Range: ____

Solve.

3. The science class took the temperature in different places around the school. Here are their temperatures in degrees Fahrenheit: 69, 61, 77, 71, 66, 68, 42, 66,72, 75, 64.
   What is the mode, median, and range of the data?

   Mode: ____ ºF Median: ____ ºF
   Range: ____ ºF

4. The science class took two more temperatures: 66º and 78º. What effect would these temperatures have on the mode, median, and range of the data?

   ____

   ____

   ____

Solve.

5. The following lists the number of pages in 7 books that Diana read: 120, 115, 98, 132, 120, 115, 135. What is the median, mode, and range of the data?

   Median: ____ Mode: ____
   Range: ____

6. Diana read one more book and added its number of pages to the data. Now the range is 38, but the mode remains the same. What could be the number of pages in Diana's new book?

   ____ pages

8.3 Explore Pictographs

Use pictograph to solve.

1. In the toad-jumping contest there were 2 Southern toads. How many toads does each ◯ stand for in the pictograph?
_____ toads

2. How many Bouncing toads were in the contest?
_____ Bouncing toads

Toads at the County Fair Toad-Jumping Contest	
Red-spotted toad	◯◯
American toad	◯◯◯◯
Bouncing toad	◯◯◯◯◯
Southern toad	◯

Each ◯ = ____

Use pictograph to solve.

3. It takes Bailey 20 minutes to walk to school. How many minutes does each ● stand for in the pictograph? _____ minutes

4. How much longer does it take Bailey to eat breakfast than to shower? _____ minutes longer

Bailey's Morning Minutes	
Breakfast	●●●
Shower	●
Get Dressed	●●●●
Walk to School	●●

Each ● = ____

Use pictograph to solve.

5. How many pounds does each ◖ stand for? _____ pounds
How can you tell?

____

6. Chen weighs 25 pounds. How can you show this in the graph?

____

Weight of Dogs Being Boarded	
Maizey	◯◯◯◖
Sally	◯◯◯◯◯
Chen	
Penny	◯◯

Each ◯ = 10 lb
Each ◖ = ____

Explore Bar Graphs

8.4

Use the bar graph to solve.

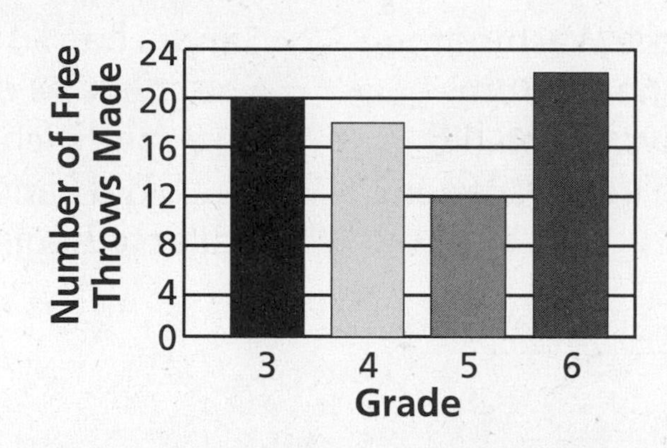

1. How many free throws did Grade 3 make on Field Day?

   ______ free throws

2. Which grade shown in the graph made the most free throws?
   Grade ______

   How can you tell?

   ____

Use the bar graph above to solve.

3. How many free throws did Grade 4 make?
   ______ free throws

   How can you tell?

   ____

4. What numbers does the scale show? ____________________

   How would the bar graph change if you changed the scale to show every two numbers?

   ____

Use the bar graph above to solve.

5. How would you change the bar graph to show that Grade 2 made 6 free throws?

   ____

6. How many free throws were made all together by Grades 3 through 6?

   ______ free throws

8.5 Problem Solving: Strategy Work Backward

Work backward to solve.

1. On March 12th, the Washington family had been home from vacation for 8 days. When did they arrive home from vacation?

____

2. Sandra has $6 left in her wallet. She spent $2 on a drink and $4 on a sandwich for lunch. How much did Sandra have in her wallet to begin with?

Work backward to solve.

3. Loni has five grades in her science class. The median is 87. The range is 12. The modes are 84 and 87. List all of Loni's science grades.

____

4. The principal had a budget for school supplies. He bought new math books for the 3rd grade for $5,585. He spent $4,090 on a scoreboard for the gym. And he spent $3,500 on new calculators. Now he has $2,000 left. How much was the total budget for school supplies?

Work backward to solve.

5. Tai joined the Hiking Club in August, five months after his 16th birthday. Three months before his birthday, he joined the Tennis Club. In what month did Tai join the Tennis club?

____

6. José lives 5 minutes from school. He walks home for lunch, then returns to school by 1:15 P.M. If José takes 20 minutes to eat lunch, what time does he leave school for lunch?

____

Coordinate Graphs

 Use the neighborhood map to solve.

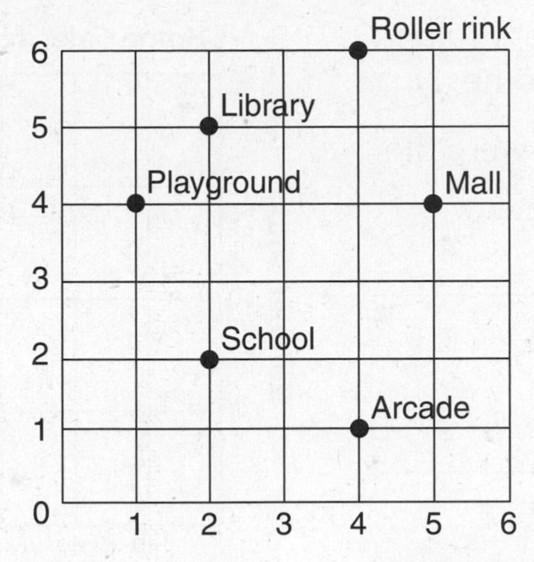

1. At what point on the map is the playground located? ______

2. What place is located at (4,1) on the map? ______________

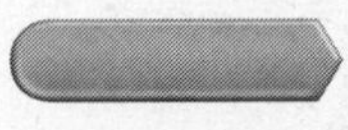 Use the neighborhood map above to solve.

3. Each space on the map stands for 1 block. How many blocks must you walk to go from School to the Arcade? ______ blocks
   At what point will you begin?
   ______
   At what point will you end?
   ______

4. If you walk two blocks to the right and 1 block up from the Playground, at what point on the map will you be? ______
   How many blocks away from the Library is this? ______________

 Use the neighborhood map above to solve.

5. If you connect the location of the Library and School with two other points not labeled on the map, you could draw a square. What are the two other points?
   ______ and ______

6. The town wants to build a bus station three blocks from the Mall. Name two possible points on the map where they might build it. ______ and ______
   How far will each point be from the School?
   ______ blocks from school

Interpret Line Graphs

Use the line graph to solve.

1. How many homes were sold in February? ______ homes

2. During which month were the most homes sold?

____

How can you tell?

____

Home Sales for First Half of 2003

Use the line graph above to solve.

3. How many homes were sold in April? ______ homes
How can you tell?

____

4. During which month were there 2 more homes sold than the number sold in May?

____

Use the line graph above to solve.

5. What is the range for the number of homes sold from January through June? ______

6. How many more homes were sold in the two months with the highest number of sales than in the two months with the least number of sales?

______ more homes
Explain. ______________________________

____

Explore the Meaning of Multiplication

Solve.

1. Paul has 3 boxes of crayons. There are 6 crayons in each box.

Use the model to tell how many crayons in all. _______ crayons

2. There are 3 children in each of the 5 reading groups.

Use the model to tell how many children in all. _______ children

Solve. Use connecting cubes if you need help.

3. Eli puts 4 books on each of 3 shelves. How many books did he put on the shelves all together?

_______ books

4. There are 8 pages in a picture book. Each page has 2 pictures. How many pictures are there in all in the book?

_______ pictures

Solve. Use connecting cubes if you need help.

5. Tim has 3 pages to fill in his trading card album. Each page holds 9 cards. Sandy has 4 pages to fill in her album, but each page holds 6 cards. Who needs more cards to fill their album? Explain.

____

____

6. There are 7 packs of apple juice and 4 packs of grape juice. Each pack has 6 boxes of juice in it. How many boxes of juice are there all together?

_______ boxes

9.2 Algebra: Relate Multiplication and Addition

Write an addition sentence and a multiplication sentence to solve.

1. There are 3 people sitting at each of 4 tables. How many people are there in all?

   ____

   ____

2. Alisa needs to put 2 forks at each of 8 table settings. How many forks in all does she need?

   ____

   ____

Solve.

3. Renee jogs 5 miles a day, 4 days each week. How many miles does she jog each week?

   ______ miles

4. Henry lives 3 miles away from the mall. Henry can run a mile in 6 minutes. If he can keep up this speed, how long will it take him to run to the mall?

   ______ minutes

Solve.

5. It takes Sam 5 minutes to wash a window. Sam has 9 windows in his house to wash. If he starts washing at 9:30 A.M., at what time will he finish?

   ____

6. Heather spent $4 for a salad and $2 for a drink. She bought the same lunch for 3 of her friends. She paid with three $10-bills. How much change did she get back?

   ____

Explore Multiplication Using Arrays

Solve.

1. Mr. Turner has 4 students in each of 5 math groups. Draw an array of circles to show how many students in all.

   _______ students

2. Four students have 3 pencils each. Draw an array of circles to show how many pencils in all.

   _______ pencils

Use the commutative property and arrays of counters to solve.

3. The top shelf in the bakery has 5 muffins on each of 6 plates. The bottom shelf has 6 muffins on each plate. Both shelves have the same number of muffins. How many plates are on the bottom shelf?

   _______ plates

   How many muffins on each shelf?

   _______ muffins

4. Each baker uses the same number of cherries. Tanya puts 3 cherries on each of 6 pies. Russell puts cherries on 3 pies. If Russell puts the same number of cherries on each pie, how many cherries does he need for each pie?

   _______ cherries

   How many cherries did each baker use?

   _______ cherries

Solve.

5. Leroy and Vern each have the same number of video games. Leroy puts an equal number of games in each of 7 boxes. Vern only has 3 boxes. He puts 7 games in each box. How many video games do Leroy and Vern have all together?

   _______ video games

6. Ray makes an array that has 4 rows of 4 counters. He wants to make two more arrays using the same number of counters. He wants more than one counter in each row. What two arrays can he make?

   ____

Problem Solving: Skill
Choose an Operation

Solve.

1. The Marching Band has 3 rows of flute players. There are 6 players in each row. How many flute players are in the band?

   _______ flute players

   Which operation did you choose and why?

   ____

2. There are 37 members in the Marching Band and 49 in the orchestra. How many are there in both groups?

   _______ members

   Which operation did you choose and why?

   ____

Solve.

3. An adult ticket to the school play costs $4.25. A child's ticket costs $2.75. How much will tickets cost for one adult and one child?

   ____

   Which operation did you choose and why?

   ____

4. A popcorn and drink combo costs $5. Mr. Harris bought 9 combos for himself and his family. How much does he pay all together?

   ____

   Which operation did you choose and why?

   ____

Solve.

5. Marla, Josh, and Guy each buy 3 new pens. The pens cost $4 each. How much did they spend in all on pens?

   ____

   Which operation did you choose and why?

   ____

6. A large box has 8 pens. The small box has 4 pens. Craig buys 3 large and 2 small boxes. How many pens did he buy?

   _______ pens

   Which operation did you choose and why?

   ____

Multiply by 2 and 5

Solve.

1. There are 2 daisies in each vase. There are 8 vases. How many daisies are there in all?

   _______ daisies

2. Maria plants 5 tomato seeds in each flower pot. If there are 6 flower pots, how many tomato seeds did Maria plant?

   _______ tomato seeds

Solve.

3. There are 7 people in the Smith family. They all keep their gloves in one box in the closet. Each person has 2 pairs of gloves. How many pairs of gloves are in the box?

   _______ pairs of gloves

4. Dad bought a new sled for the children to share. He paid with eight $5-bills. He did not get any change back. How much did the sled cost?

   ____

Solve.

5. Letti is coloring 9 flowers on a page. Each flower has 5 petals. So far Letti has colored in 28 petals. How many more petals must she color to finish?

   _______ more petals

6. Mrs. Ortiz bought 9 coloring books as party favors. The books are $2 each. She paid with a $20 bill. How much change should Mrs. Ortiz get back?

   ____

10.2 Problem Solving: Strategy Find a Pattern

Find a pattern to solve.

1. There are 6 rows of vegetables in the garden. Row 1 has 2 vegetables, Row 2 has 4 vegetables, and Row 3 has 6 vegetables. If the rest of the garden follows this pattern, how many vegetables are in Row 6?

   _______ vegetables

2. The children in the art club are standing in line to get art supplies. Every 5th child will get an art box to bring to the group table. The rest will get paintbrushes for the group. What will the 10th child in line get?

   ____

Find a pattern to solve.

3. Keenan just started a jogging program. He must jog 4 miles the first week, 8 miles the second week, and 12 miles the third week. If he continues in the same pattern for two more weeks, how many miles will he jog the fifth week?

   _______ miles

4. The Mayfair Softball team colors are red and white. The team stood in line for a team picture. Every second player wears a red cap. The rest wear white caps. What color cap is the first player in line wearing?

   ____

Find a pattern to solve.

5. There are 35 stars around the border of a bulletin board display. Every fourth star is blue. The rest are gold. How many blue stars are around the border?

   _______ blue stars

6. Trent needs $20 to buy a CD. He saves a total of $3 by the end of week 1, $6 by the end of week 2, and $9 by the end of week 3. If he keeps saving in the same way, how many weeks will it take for him to save enough money to buy the CD?

   _______ weeks

Multiply by 3 and 4

Solve.

1. The straight part of Eli's train track has 4 tracks. Each track is 7 inches. How many inches long is the straight part of the train track?

   _______ inches

2. Melissa owns 3 sets of trains. Each set has 6 train cars. How many train cars does Melissa have in all?

   _______ cars

Solve.

3. The border around a bulletin board is 35 inches long. There are 4 pieces of border paper left. Each piece is 9 inches long. Is there enough border paper to go around the bulletin board border? Explain.

   ____

4. There are 3 groups of students in charge of decorating the hallway bulletin boards. Each group decorates 8 different boards around the school. How many bulletin boards are there in all?

   _______ bulletin boards

Solve.

5. Paula can make 4 beaded bracelets in an hour. Tina can make 3 beaded bracelets in an hour. In one week Paula made bracelets for 6 hours and Tina made them for 8 hours. Who makes more bracelets in the week? Explain.

   ____

6. Each box has 30 of the same colored beads. Every bracelet has 4 blue beads and 3 red beads. If Jackie makes 5 bracelets, how many beads will be left in the box of blue beads?

   _______ blue beads

   How many will be left in the box of red beads?

   _______ red beads

Algebra: Multiply by 0 and 1

Solve.

1. There are 10 people sitting in 1 row of the concert hall. How many people are in this row?

_______ people

2. When the concert is over there are 0 people in row 1. How many people are in this row?

_______ people

Solve.

3. If 0 people paid $5 each for a ticket to a school play, how much did they spend in all?

____

4. There are 6 chairs at 1 table in the dining room. How many chairs are there in the dining room?

_______ chairs

Solve.

5. Beth put 4 stickers on each of 3 white pages in her album. She put 8 stickers on each of 2 yellow pages, and 0 stickers on each of 7 red pages. How many stickers did she put in her album in all?

_______ stickers

6. There are 4 bowls of fruit on a table. Each bowl has 3 apples, 2 bananas, and 1 orange. How many oranges are there in all?

_______ oranges

Explore Square Numbers

Solve.

1. The model shows the square tiles on a floor.

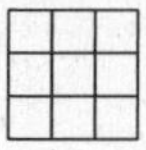

   Write a multiplication sentence to show the number of tiles in all.

   ____

   How many tiles in all?

   ______ tiles in all

2. There are 2 rows of square tiles and 2 tiles in each row. How many square tiles are in a floor? Use a model to solve.

   ______ square tiles

Solve.

3. Marvin bought ceiling tiles for his square basement. The ceiling is 8 square tiles long. How many tiles will he need to cover the whole ceiling?

   ______ tiles

4. What square number is less than 64 but greater than 36? Write a multiplication sentence to solve.

   ____

Solve.

5. Pat made a quilted pillow using squares of material. The pillow has 4 rows of 3 squares each. Did Pat make a square pillow? Why or why not?

   ____

   ____

6. The model shows a square floor with some square tiles.

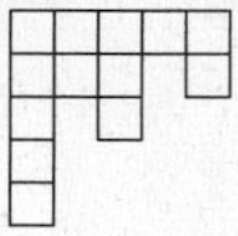

   How many more tiles must be put down to cover the floor?

   ______ more tiles

11.2 Multiply by 6 and 8

Solve.

1. A river raft holds 8 people. There are 2 rafts for rent. How many people can raft along the river at one time?

   _______ people

2. Mark has 6 friends in each of 3 different states. How many out-of-state friends does he have in all?

   _______ friends

Solve.

3. Mickie has 8 nickels. How much money does she have?

   ____

4. Sonia scored 9 points in each of the last 6 basketball games. How many points has she scored so far for the season?

   _______ points

Solve.

5. Sara pays for 7 new pairs of socks with five $10 bills. Each pair of socks costs $6. Can Sara buy a new $4 scarf with her change? Explain.

   ____

   ____

6. Bob brought his sports card collection over to Tyrell's house. Bob keeps his cards in an album that has 8 pages. Each page holds 6 cards. When he arrived, he realized that he had left two pages at home. How many cards did Bob bring to Tyrell's house?

   _______ cards

Multiply by 7

Solve.

1. The Martins will buy 2 new tires for all 7 of their bicycles. How many new tires will they buy?

   _______ new tires

2. It takes Cally 3 minutes to paint each slat on a fence. There are 7 slats in each section of the fence. How long will it take Cally to paint each section of the fence?

   _______ minutes

Solve.

3. Each house on Alpine Street has 7 front windows. There are 3 houses on each side of the street. How many front windows are there in all?

   _______ windows

4. Mario will go on vacation for 8 weeks this summer. For how many days will Mario be on vacation?

   _______ days

Solve.

5. Nell bought 3 pairs of white socks and 4 pairs of black socks. Each pair cost $6. Then she bought a $5.75 hat. She got back $12.25 in change. How much did Nell give to the cashier to pay for the socks and hat? Show your work.

   ____

   ____

6. There are an equal number of cars and bicycles in the garage. If there are 42 tires in all, how many bicycles and cars are in the garage? Explain.

   ____

   ____

11.4 Algebra: Find Missing Factors

Solve. List the fact family if you need help.

1. Joe can read a page in 3 minutes. How many pages can he read in 6 minutes?

   ______ pages

2. Mel read the first 3 lessons of his book. Each lesson has the same number of pages. Mel read 15 pages. How many pages are in each lesson?

   ______ pages

Find missing factors to solve.

3. Each display shelf has 5 display plates. There are 35 plates in all. How many shelves are there?

   ______ $\times 5 = 35$

   ______ shelves

4. Ben has 45 cents in his pocket. All of the coins are nickels. How many nickels does Ben have in his pocket?

   ______ $\times 5 = 45$

   ______ nickels

Find missing factors to solve.

5. Mr. Lowry paid for 8 same-priced shirts with a \$100-bill. He got back \$28 in change and used that to buy 4 same-priced ties. Now he has no money left. How much did each shirt and each tie cost?

   ____

   ____

6. Hector did 16 push-ups in 2 minutes. Ralph did 27 push-ups in 3 minutes. Who can do push-ups faster, Hector or Ralph? Explain.

   ____

   ____

Problem Solving: Strategy
Solve Multistep Problems

Solve.

1. In a roller coaster car, there are 3 people in each of the first 2 rows. There are 2 more people in the last row. How many people are in the car?

   _______ people

2. On the carousel, the bench has 6 children on it. 5 of the horses have 2 children each. How many children are on the carousel?

   _______ children

Solve.

3. Alan lives 4 blocks from school. He walks to and from school each day. How many blocks does he walk in all each school week?

   _______ blocks

4. Small bottles of juice cost $2, and large bottles cost $5. How much does it cost to buy 6 bottles of each size?

   ____

Solve.

5. Ms. Wallace has $25 to take her 2 children to lunch. She buys each child a sandwich for $4.95 and a drink for $1.95. She buys her own lunch for $7. If she leaves a $4 tip for the waitress, how much money did she have left?

   ____

6. Roy and Len wash windows on skyscrapers. Roy can wash 8 windows in an hour. Len can wash 6 windows in an hour. Each worker washed 48 windows today. How much longer did Len work than Roy? Explain.

   ____

   ____

12.1 Multiply by 10

Solve.

1. In a game, Carlos ran with the football three times. Each time, he ran 10 yards. How many yards did he run?

_______ yards

2. The Appletown Zoo has 10 monkeys. Each monkey gets one banana a day. How many bananas do the monkeys eat each day?

_______ bananas

Solve.

3. Hal shoes horses on a farm. Today he put horseshoes on all the hooves of 10 horses. How many horseshoes did he put on?

_______ horseshoes

4. There are 9 women who have an appointment at the nail salon. Kiki will polish all of their fingernails. How many fingernails will Kiki polish today?

_______ fingernails

Solve.

5. A children's TV show is on 10 days each month. On every show, Burton the clown plays 3 songs. In the last month, he sang six of all the songs that were played. How many times did he *not* sing last month?

_______ times

6. Ellen drives a van for the animal shelter. The van can hold up to 10 animals. This week she made 6 trips to the shelter. On 4 of those trips, the van was filled with dogs. On the rest of the trips, the van was filled with cats. How many more dogs than cats did she bring to the shelter this week?

_______ more dogs

Multiply by 9

Solve.

1. Jose spends \$9 on lunch each day. How much does he spend for lunch in 2 days?

____

2. Carmen's parrot eats 9 crackers a day. How many crackers will it eat in 4 days?

_______ crackers

Solve.

3. On Mr. Dugan's farm, 9 cows can be milked in an hour. Mr. Dugan says that 45 cows will be milked in 5 hours. Is he correct? Explain.

____

4. The So Rich cookie factory can bake 9 chocolate chip cookies a minute. Can the factory fill an order for 80 cookies in 9 minutes? Explain.

____

Solve.

5. For the school talent contest, 9 singers will perform for 3 minutes each. Then 5 dancers will perform for 4 minutes each. There will be a 15-minute intermission between the singing and dancing performances. If the show starts at 2:00 P.M., at what time will it end?

____

6. Ty works 9 hours a day and earns \$6 an hour. Cal works 6 hours and day and earns \$9 an hour. If they both work 5 days per week, who earns more money?

____

Who works longer? Explain.

____

____

12.3 Multiplication Table

Use the multiplication table to solve.

1. There are 2 cashiers in each department in a store. There are 11 departments. How many cashiers work in the store?

   _______ cashiers

2. Andrea is buying a new dress. She wants to try on 12 different dresses. It takes her 3 minutes to try on each dress. How long will it take to try on all of the dresses?

   _______ minutes

Use the multiplication table to solve.

3. There are 12 items in a dozen. Each box of pencils has a dozen pencils. How many pencils are there in 12 dozen boxes?

   _______ pencils

4. Kit is 11 years old. Grandma is 6 times as old as Kit. How old is Grandma?

   _______ years old

Solve.

5. Luke rides 12 miles in one hour on his bike. He is riding in a race that is 31 miles. So far, he has biked for 2 hours. Does he have *more than* an hour or *less than* an hour to complete the race? Explain.

   ____

   ____

6. Travis is 6 ft tall. Mt. McKinley, in Alaska, is 20,254 ft plus 11 times as tall as Travis. How high is Mt. McKinley?

   _______ ft

Algebra: Multiply 3 Numbers

Solve.

1. There are 3 rows of benches in a rest area on the boardwalk. Each row has 3 benches. If 3 people can fit in each bench, how many people can sit in the rest area at once?

   _______ people

2. Shari has some seashells on shelves in her room. She has 4 shells in each box with 2 boxes on each shelf. If there are 2 shelves on the wall, how many sea shells are there in all?

   _______ sea shells

Solve.

3. Mr. Moro gives 2 dimes to each of his 3 children. Write a multiplication sentence to show how much money in pennies Mr. Moro gave his children.

   ____

4. A parking lot charges $5 per car. There are 10 parking spaces in each row. If 2 rows of spaces are full, how much money will the parking lot attendant collect? Write a multiplication sentence.

   ____

Solve.

5. The classroom has 5 rows of desks with 4 desks in each row. Each desk has 3 books in it. So far, the students have covered 25 books. How many more books need to be covered?

   _______ more

6. Stacey and Anna write their first names in cursive on a sheet of paper with 2 columns. They write their names 10 times in each column. How many more letters does Stacey write than Anna does? Explain.

   ____

   ____

12.5

Problem Solving: Strategy Draw A Picture

Draw a picture to solve. Use a separate sheet of paper.

1. Meg wears rings on 3 of her fingers on each hand. She has 2 rings on each of those fingers. How many rings does she wear in all?

   _______ rings

2. There are 2 shelves in a store window. Each shelf has 5 necklaces. Each necklace has 10 pearls. How many pearls in all?

   _______ pearls

Draw a picture to solve. Use a separate sheet of paper.

3. A calculator has 7 rows of keys. Each row has 4 keys and each key has 2 symbols on it. How many symbols in all?

   _______ symbols

4. Carly, Emma, and Grace each buy a white shirt and a blue shirt. The shirts cost $9 each. What is the total cost of all the shirts?

   ____

Draw a picture to solve. Use a separate sheet of paper.

5. Three friends each have four $5 bills in each of their two pants pockets. They put their money together to buy a set of tools to share. The tool set costs $131.99. How much more do they need to buy the tool set?

   _______ more

6. Heather decorated her wall by painting 6 big squares with 2 circles in each square and 3 stars in each circle. Haley painted 8 big circles on her wall with 4 triangles in each circle and 2 stars in each triangle. Who has more stars on her wall?

   ____

   How many more stars?

   _______ more stars

Explore the Meaning of Division

Use models to help you divide.

1. Angel has 2 boxes of markers. Each box has the same number of markers. There are 12 markers in all.

   How many markers are in each box?

   _______ markers

2. There are 8 children at 4 tables. The same number of children are at each table.

   How many children are at each table?

   _______ children

Use models to divide.

3. Russ, Mike, and Tom carry the same number of books in each of their backpacks. They carry 21 books in all. How many books are in each backpack?

   _______ books

4. There are 4 drawers in a dresser. Each drawer has an equal number of handles on it. If 6 of the handles are brass and 6 are copper, how many handles are on each dresser drawer?

   _______ handles

Divide. Use models if you need help.

5. The hardware store is selling 3 boxes of Double-A batteries for $12. There are 24 batteries all together in the boxes. How much does 1 box of batteries cost?

   _____

   How many batteries are there in 1 box?

   _______ batteries

6. The art teacher will give 8 students in the art club the same number of paint brushes. She has 35 paintbrushes. What is the greatest number of paint brushes each student can get? Explain.

   ____

   ____

Algebra: Relate Division and Subtraction

Use subtraction to solve.

1. Perry puts 9 strawberries into 3 fruit cups. He puts the same number of strawberries in each cup. Use subtraction to show how many strawberries he put in each cup.

   ____

   How many strawberries did he put in each cup?

   ______ strawberries

2. Four people at the Pizza Palace left the waiter a tip. Each person left a tip of the same amount. The total tip was $8. Use subtraction to show how much each person left for a tip.

   ____

   How much did each person leave for a tip?

   ____

Use subtraction to solve.

3. On Monday, Helen's math teacher gave the class 45 problems to finish by Friday. Helen will do the same number of problems each day. How many problems will she do on Friday?

   ______ problems

4. The school cafeteria can serve lunch to 4 students every 32 seconds. How many seconds does it take each student to get his or her lunch?

   ______ seconds

Use subtraction to solve.

5. A box of tissue packs contains 72 total tissues. The tissues come in packs of 8 tissues each. Ally, Ann, and Missy share the tissue packs equally. How many packs of tissues does each girl get? Explain.

   ____

   ____

6. Four friends buy tickets to see a movie. They pay $24 in all for their tickets. If each friend also spends $2 on a drink, how much does each friend spend in all? Explain.

   ____

   ____

Algebra: Relate Multiplication to Division

Draw a picture to solve. Use a separate sheet of paper.

1. A mini-van has 3 rows of seats with 9 seats in all. Draw an array of circles to show the number of seats in each row. How many seats in each row?

   _______ seats in each row

2. Two students have 10 pennies in all. They each have the same number of pennies. Draw an array of circles to show how many pennies each student has. How many does each have?

   Each has _____ pennies

Use arrays of counters to help you solve.

3. A news reporter spent the last 24 months in 6 different countries. She stayed the same length of time in each country. How long did she stay in one country?

   _______ months

4. The news channel on TV allows 20 minutes to report the day's top stories. Today's top stories took 5 minutes each to report. How many top stories were reported today?

   _______ top stories

Solve. Use arrays if you need help.

5. Nina made 6 pairs of pants with 42 pockets in all. Each pair of pants has the same number of pockets. She added a button to one pocket on each pair of pants. How many pockets on each pair of pants do **not** have buttons?

   _______ pockets

6. The math teacher gives Harlen 24 counters. Harlen must make as many different arrays as he can with more than 1 row. How many different arrays can he make? [Remember: In an array each row has the same number of counters]

   _______ different arrays

   List the arrays.

   ____

Divide by 2

Solve.

1. Britt spent the same amount of money at 2 different stores. She spent $2 in all. How many groups of 2 are there in $2?

   ____

   How much did she spend at each store?

   ____

2. Tyrell gave 4 of his model cars to his friends Ted and Ameil. He gave the same number of cars to each friend. Write a division fact to show how many cars Tyrell gave to Ted.

   ____

   How many cars did he give to Ted?

   ______ cars

Solve.

3. Casey bought a box of 18 granola bars. She will keep some and give the rest to her brother. If Casey and her brother now have the same number of bars, how many did Casey give to him?

   ______ bars

4. Mother washes all 14 of her children's mittens. Each child has one pair of mittens. How many children are there?

   ______ children

Solve.

5. Jodie is helping her mom in the backyard. She needs to move 17 big stones to the front. The wheelbarrow can hold 2 stones. Can she move all of the stones to the front yard in 8 trips? Explain.

   ____

   ____

6. Ian is cleaning his room. He picked up 16 red pegs and 12 black ones. He put the same number of each color into each of two boxes. How many pegs did he put in each box?

   ______ pegs in each box

Problem Solving: Skill
Choose an Operation

Solve. Circle the operation you used.

1. Flora has 8 pennies in one hand. She has the same number in her other hand. How many pennies does she have in all?

   _______ pennies

   Addition Subtraction
   Multiplication Division

2. Carmine fixed 4 bicycles. It took him 12 hours to fix them. He worked the same number of hours on each bike. How many hours did it take to fix one bike?

   _______ hours

   Addition Subtraction
   Multiplication Division

Solve. Write the operation you used.

3. Uri is making a 7-layer cake. Each layer will need 3 ounces of pudding. How much pudding will Uri need to make the cake?

   _______ ounces

   Operation: _______________

4. The town of Newton received 27 inches of snow. If three inches of snow melt each day, how long will it take for all of the snow to melt?

   _______ days

   Operation: _______________

Solve. Write the operations you used.

5. A baker's closet has 3 small shelves and 5 large shelves. There are 4 jars of peanut butter on each of the small shelves and 6 jars on each of the large shelves. How many jars in all are in the closet?

   _______ jars

   Operations: _______________

6. Gloria is reading a book for a report. She has read for 35 minutes so far. She reads one lesson every 5 minutes. If Gloria decides to read another 3 lessons, how many lessons will she have read in all?

   _______ lessons

   Operations: _______________

Divide by 5

Solve.

1. Antonio scored 15 points on 5 math questions on a test. Each question was worth the same number of points. How many points did he score for each question?

   _______ points

2. School lunch costs $5. Marcus has $10. For how many days can he buy lunch?

   _______ days

Solve.

3. Erica works at a pet store. It takes her five minutes to put food and water in each hamster cage. How many cages can she finish in 35 minutes?

   _______ cages

4. Joel is in charge of feeding the birds in a pet store. Each bird cage gets 5 hanging seed strings. Joel used 45 seed strings to feed all of the birds. How many cages of birds are in the store?

   _______ cages

Solve. Show your work.

5. Every Saturday, Mr. and Mrs. Thompson and their 3 children each have a hamburger for lunch. There are 40 hamburger patties in their freezer. In how many weeks will they finish the last of the patties?

   ____

   ____

6. Today 25 girls and 20 boys rode their bikes to school. Each bike rack at school holds 5 bikes. How many bike racks were filled?

   ____

   ____

Divide by 3

Solve.

1. Walter has 9 pencils. Every week he uses 3 of them. In how many weeks will Walter use up all of his pencils?

   _______ weeks

2. Alana mailed 6 letters in 3 different mailboxes. She mailed the same number of letters in each mailbox. How many letters did she mail in each mailbox?

   _______ letters

Solve.

3. Elyse served herself and 2 friends 24 ounces of juice. She filled each glass with the same amount of juice. How many ounces of juice did she pour in each glass?

   _______ ounces

4. The 27 students in Mrs. Penny's class are in line to leave school. Mrs. Penny lets her students leave in groups of 3 at a time. How many groups of students will leave?

   _______ groups

Solve. Show your work.

5. The gym teacher has 18 basketballs divided equally among 3 bags. For practice she takes 2 basketballs from each bag. How many basketballs are left in one of the bags?

   ____

   ____

6. All three of Tasha's dogs eat the same amount of food . She feeds them a total of 12 pounds of dry food and 12 pounds of canned food every week. How many pounds of food does each dog eat per week?

   ____

   ____

14.3

Divide by 4

Solve.

1. Each minute, 4 gallons of water flow into the tub. There are now 8 gallons of water in the tub. How many minutes did that take?

   ________ minutes

2. The Finos have a carton of 12 eggs. If the family eats four eggs a day, how long will they have eggs to eat?

   ________ days

Solve.

3. Eric pumps the front tire of his bike to 32 pounds. Each push of the pump puts 4 pounds of air into the tire. How many times must Eric push the pump to fill the tire?

   ________ times

4. A boat rental shop rents paddleboats that can hold up to 4 riders. The shop has enough paddle boats for up to 28 people. How many paddleboats does the shop have?

   ________ paddleboats

Solve.

5. Ollie lent $24 in equal amounts to 4 of his friends. Melissa lent $18 in equal amounts to 3 of her friends. Who lent each friend more money? Explain.

   ____

   ____

6. A grocery store shelf can hold 4 large boxes of Pride detergent. The store clerk put 25 boxes of Pride on the shelves. What is the least number of shelves needed for the display? Explain.

   ____

   ____

Algebra: Divide with 0 and 1

Solve.

1. Kelly divided 0 shirts into 4 equal groups. How many shirts are in each group?

_______ shirts

2. A delivery man carries 5 new chairs into 5 rooms. He puts the same number of chairs in each room. How many chairs are in each room?

_______ chair(s)

Solve.

3. Each desk in an office has 1 chair. There are 8 chairs in all in the office. Write a number sentence to show how many desks are in the office.

____

How many desks are in the office?

_______ desks

4. Mandy arranged pictures of her family in 3 equal rows on her wall. Mandy has 3 pictures of her family. How many pictures are in each row?

____

Solve. Show your work.

5. A florist has 32 daisies and 8 lilies to arrange in 8 vases. She puts the same number of each kind of flower in each vase. She then puts another 5 rosebuds in each vase. How many flowers in all are in each vase?

____

____

6. A gardener plants 18 tulips in 6 equal rows. If she does the same with 0 marigolds and 24 daffodils, how many marigolds will be in each row?

____

How many more daffodils than tulips will be in each row?

____

14.5 Problem Solving: Strategy Act It Out

Act out the problems to solve.

1. Erin put 3 stickers on each picture she made. She used 6 stickers in all. How many pictures did Erin make?

   _______ pictures

2. Each of 4 children is wearing 4 bangle bracelets. How many bracelets are there in all?

   _______ bracelets

Act out the problems to solve.

3. Each pack of printing paper costs $4. How much do 9 packs of paper cost?

   ____

4. Greeting cards come in boxes of 32 cards. A mixed box includes equal amounts of thank you cards, birthday cards, friendship cards, and anniversary cards. How many birthday cards are there?

   _______ birthday cards

Solve. Act out the problems if you need help.

5. Each of 5 families has 3 tricycles. All of the tricycles have streamers on both handlebars. If two of the families take the streamers off the handlebars, how many streamers are still on the tricycles?

   ____

6. Each wheel on Jason's bicycle has 8 spokes. Every spoke has 1 reflector. Jason takes off 3 reflectors from one wheel. How many reflectors are left on his bike?

   _______ reflectors

Divide by 6 and 7

Solve.

1. Len will put 18 goldfish into 6 fishbowls. Each bowl will have the same number of fish. How many goldfish will go in each bowl?

   _______ goldfish

2. There are 14 customers standing in 7 checkout lines. Each line has the same number of customers. How many customers are in each line?

   _______ customers

Solve.

3. There are 54 cards in a card game. All of the cards are dealt out to the players. Each player gets 6 cards. How many players are in the game?

   _______ players

4. The winning team scored 49 points. There were 7 players on the team. If each player scored the same number of points, how many points did each player score?

   _______ points

Solve.

5. Mother is making 6 goody bags for Leroy's party. She will put 24 apple fruit rolls and 24 cherry fruit rolls into the bags. If she puts the same number in each bag, how many fruit rolls will be in each goody bag?

   _______ fruit rolls

6. There are 7 cupcakes for the party. Each cupcake has 1 candle for each year of the birthday boy's age. There is also an extra candle on each cupcake for good luck. If 49 candles were used on the cupcakes, how old is the birthday boy? Explain.

   ____

   ____

Problem Solving: Skill
Solve Multistep Problems

Solve multistep problems.

1. One child has 2 boxes of crayons with 8 crayons in each box. Another has 1 box of 8 crayons. How many crayons in all?

   _______ crayons

   They put all of the crayons together and divided them equally into 6 little bags. How many crayons in each bag?

   _______ crayons

2. Tracey pasted 4 stars on each of her 3 pictures. Alan pasted 3 stars on each of his 6 pictures. How many stars did Alan paste?

   _______ stars

   How many more stars did Alan paste than Tracey?

   _______ more

Solve multistep problems.

3. Jim has 32 pennies. Rick has 8 fewer pennies than Jim. They combine their money and use it to buy 7 cookies at a bake sale. If each cookie costs the same price, how much is one cookie?

   _______ cents

4. Hanna talked on the phone for 5 minutes every day for a week. Her brother talked on the phone for 8 minutes each day. How many minutes in all did they talk on the phone for the week?

   _______ minutes

Solve. Explain your answers.

5. Sal exercised for 12 minutes one day. He doubled his exercise time on the second day. He will exercise the same number of minutes on the third day as he did on the first 2 days together. He will do each exercise for 9 minutes. How many exercises will Sal do on the third day? Explain.

   ____

6. There are 24 extra tickets for the game. They will be divided equally among the 3 boys and 3 girls on the basketball team. If each player pays a total of $20 for tickets, how much does each ticket cost? Explain.

   ____

Divide by 8 and 9

Solve.

1. A group of 8 children go to the fair. They share 16 balloons equally. How many balloons does each child get?

   _______ balloons

2. A group of 9 people go on 27 rides at the fair. Each one goes on the same number of rides. How many rides does each person go on?

   _______ rides

Solve.

3. Marta bought 48 pieces of flatware. She puts them in a tray with 8 sections. Each section has the same number of pieces. How many pieces of flatware are in each section of the tray?

   _______ pieces

4. Mina sets the dining room table. Every night she puts out 45 dishes. She sets 9 places at the table. How many pieces of dinnerware are there at each place?

   _______ pieces

Solve.

5. Ty and Shaheed each have 36 rocks. They put their rocks together in a box. The box has 9 sections. If they put the same number of rocks in each section, how many rocks are in each? Explain.

   ____

   ____

6. A mural in the aquarium shows octopuses and starfish. Each starfish has 5 arms. Each octopus has 8 legs. There are 20 starfish arms in all. The combined number of starfish arms and octopus legs is 60. How many octopuses are in the mural? Explain.

   ____

15.4 Explore Dividing by 10

Solve.

1. There are 30 desks with 10 desks in each row. How many rows of desks are there?

   ________ rows

2. Carl owns 20 video games. He stores them in boxes. There are 10 video games in each box. How many boxes are there?

   ________ boxes

Solve.

3. Mary kept a record for 90 days to see how many times she ate fish for dinner. She ate fish every 10 days. How many times did she have fish for dinner in the last 90 days?

   ________ times

4. Annie bought a bag of 80 mini-carrots. She eats 5 carrots each day for lunch and eats another 5 carrots as a snack at night. In how many days will the bag of carrots be empty?

   ________ days

Solve.

5. Morgan has 90 cents in her pocket and another 80 cents in her drawer. All of the change is in dimes. How many dimes does Morgan have in all?

   ________ dimes

6. Ricky spent $95 at the supermarket. He bought $25 worth of fruit. The rest of the money was spent on steaks. If he bought 10 steaks and each cost the same amount, what was the price of each steak?

   ____

Use a Multiplication Table

Use the multiplication table to solve.

1. There are 36 steps to climb. There are 12 steps on each flight. How many flights of steps are there to climb?

   _______ flights of steps

2. Ted painted 11 of the steps in 22 minutes. How long did it take to paint each step?

   _______ minutes

Use the multiplication table to solve.

3. There are 12 items in a dozen. Each carton of eggs holds a dozen. The school cafeteria has 84 eggs. How many cartons of eggs are there?

   _______ cartons

4. The baseball coach paid $132 for baseball caps for the players on the team. There are 11 players in all. How much did each baseball cap cost?

   ____

Solve.

5. At a car dealership, there are 102 new cars on the lot and 6 cars in the showroom. The dealership wants to sell all of these cars by the end of the month. If each of the dozen salespeople sell the same number of cars, how many must each person sell? Explain.

   ____

6. A car salesperson wants to sell 130 cars in one year. She sold 9 cars in the first month. How many cars per month must she sell for the rest of the year to reach her goal? Explain.

   ____

16.1 Algebra: Use Related Facts

Solve. Tell which fact family helps to solve each problem.

1. Tad will deal a total of 12 cards to 3 players. Each player will get the same number of cards. How many cards will each player get?

   ______ cards

   Fact Family:

   ____

2. If 10 cards are on a table in 2 equal rows, how many cards are in each row?

   ______ cards

   Fact Family:

   ____

Solve. Show how you can use either division or multiplication.

3. There are 6 shelves in a china closet. There are 42 display plates divided equally among the shelves. How many plates are on each shelf?

   ______ plates

   Division: ______________

   Multiplication: ______________

4. Shawna pays $63 for 7 display plates. Each plate costs the same price. How much is one display plate?

   ____

   Division: ______________

   Multiplication: ______________

Solve. Show the fact families you can use to solve.

5. The school band has 24 members. Describe 3 different arrangements in which the players can march in 3 or more equal rows. There must be at least 3 players in each row.

   ____

   ____

   Fact Families: ______________

   ____

6. Vera wants to arrange 48 photographs on an album page in equal rows. She wants at least 4 rows with 4 or more photos in each. What are the possible ways she can arrange the photos?

   ____

   ____

   Fact Families: ______________

   ____

Problem Solving: Strategy Guess and Check

Use Guess and Check to solve.

1. Mrs. Marcus will sew 10 buttons. She will sew 2 buttons on the red shirts and 3 buttons on the blue shirts. There is a total of 4 shirts. How many shirts are red?

   _______ red

   How many are blue?

   _______ blue

2. Mr. O'Hara will change all 8 light bulbs in 5 lamps. Floor lamps have 2 bulbs and table lamps have 1 bulb. How many of the lamps are floor lamps?

   _______ floor lamps

   How many are table lamps?

   _______ table lamps

Use Guess and Check to solve.

3. Susan bought a fruit pop that cost $0.30. She paid for it with dimes and nickels. If she used 4 coins to pay for the pop, how many were dimes?

   _______ dimes

   How many were nickels?

   _______ nickels

4. Alice makes a collage of 9 daisies. She has 51 petals. She pastes 5 on the yellow daisies and 6 on the purple daisies. How many yellow daisies does she make?

   _______ yellow

   How many purple?

   _______ purple

Solve.

5. Ken has 7 coins. Their total value is $0.60. What coins might Ken have?

   ____

   How many of each?

   ____

6. A jeweler made 7 pearl necklaces. Some have 5 pearls and the rest have 8 pearls. She used a total of 47 pearls on the necklaces. How many more pearls did she use on the 8-pearl necklaces than on the 5-pearl necklaces?

   _______ more pearls

Explore the Mean

Use connecting cubes to solve.

1. Allan hit 3 home runs in one game and 5 home runs in another.

What is his mean number of home runs for the two games?

_______ home runs

2. Jan struck out the following number of players in her last 3 baseball games: 2, 5, 2.

What is the mean number of players Jan struck out in her last three games?

_______ players

Use connecting cubes to solve.

3. There are 3 children in the Raos family. Their ages are 3, 5, and 10. What is the mean age of the Raos children?

_______ years of age

4. Tasha's tomato plants are just starting to grow. So far their heights are 4 inches, 2 inches, 5 inches, and 5 inches. What is the mean height of Tasha's plants?

_______ inches

Solve. Use connecting cubes if you need.

5. There are five different math groups in the third grade. The following number of students are in each group: 4, 6, 5, 8, 7. What is the mean number of students in the math groups?

_______ students

6. Zoe ran 5 days in a row. On the first day she ran 6 miles. On the rest of the days, she ran one mile more than she had the day before. What is the mean number of miles Zoe ran for all five days?

_______ miles

Find the Mean

Solve.

1. In the last three days, Paul saved $5, $8, and $2. What is the mean amount of money he saved each day?

   ____

2. Four students read the following numbers of books in the past month: 8, 12, 6, and 10. What is the mean number of books read?

   ________ books

Use the data in the table to solve.

Runner	Number of Minutes Needed to Run 1 Mile	Number of Miles Run Per Day
Jara	9	3
Tricia	8	5
Devon	7	4
Scott	8	12

3. What is the mean number of miles run per day?

   ________ miles per day

4. What is the mean number of minutes it takes to run 1 mile?

   ________ minutes per mile

Solve.

5. Eight students answered 10 questions on a test. Here are the number of questions each answered correctly: 10, 8, 9, 8, 0, 10, 4, 7. If students get 10 points for each correct answer, what is the mean test score?

   ____

6. Each student in art class has made the following number of clay dinosaurs: 10, 4, 6, 9, 6, and 7. What is the mean number of dinosaurs made?

   ________ dinosaurs

   A new student joins the class. He has made 0 dinosaurs. Does the mean number of dinosaurs change? If so, how?

   ____

Explore Multiplying Multiples of 10

Use models to solve.

1. Nathan earns $30 a week at his part-time job. How much does he earn in 3 weeks?

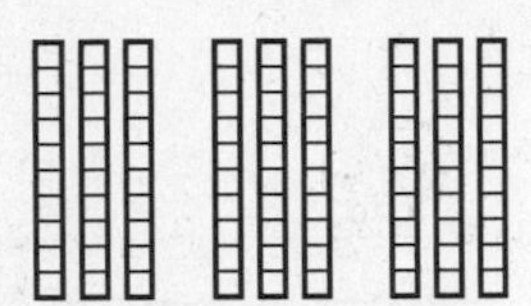

____

2. The fruit store has 2 crates of apples left to sell. There are 50 apples in each crate. How many apples are left in all?

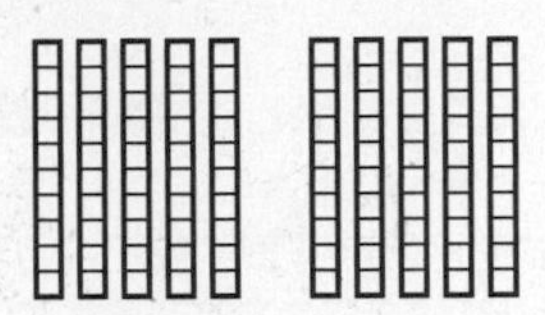

_______ apples

Use models to solve.

3. Shelley wants to make 80 copies of the third-grade class newsletter. The newsletter is 6 pages long. How many sheets of paper will she need to make the copies?

_______ sheets

4. Mr. Muscle can lift 6 boxes that weigh 40 pounds each. How much total weight can he lift?

_______ pounds

Solve. Use models if you need help.

5. Andrea earned $80 per week for the 8 weeks that she worked over the summer. She saved $45 and spent the rest on books and a vacation. How much did she spend on books and a vacation?

____

6. A carpenter made 90 new shelves. The materials for each bookshelf cost $9. He sells the shelves for a total of $1,800. How much profit did he make?

____

Algebra: Multiplication Patterns

Solve.

1. Nadia's mom drove 60 miles an hour for 4 hours. How many miles did she drive?

   ________ miles

2. Thomas has 4 boxes of model train tracks. Each box holds 20 feet of track. How many feet of track does he have in all?

   ________ feet

Solve.

3. The 6 elementary schools in Thornton School District have 300 students each. How many elementary students attend Thornton School District?

   ________ students

4. Some computers send information at the speed of 200 megabytes every second. How many megabytes could be sent in 8 seconds?

   ________ megabytes

Solve.

5. Benson School has 3 third-grade classrooms. There are 3 computers in each of the classrooms. Each computer costs $2,000. How much did all of the third-grade computers cost?

   ____

6. Robbie's dad paid $20,000 for his brand new pickup truck. Every year for the first 4 years, the truck loses $2,000 in value. What is the value of the truck at the end of the 4 years?

   ____

17.3 Estimate Products

Estimate to solve.

1. In the year 2002, Abby had 29 rocks in her collection. Now she has 4 times that number. About how many rocks does she have now?

   about _____ rocks

2. If a woodpecker can peck 19 times each second, about how many times can it peck in 3 seconds?

   about _____ times

Estimate to solve.

3. A cat can move at a speed of about 792 feet each minute. About how far can it move in 9 minutes?

   about _____ feet

4. Fred drives at an average speed of 58 miles per hour. It takes him 7 hours to drive to Florida from his home. About how many miles from Florida does Fred live?

   about _____ miles

Estimate to solve.

5. There are 110 floor tiles in a box. Each tile will cover 2 square feet. About how many boxes will you need to cover 980 square feet? Explain.

   ____

   ____

6. Dan starts delivering mail at 8:30 A.M. He usually delivers about 215 pieces of mail each hour. Dan takes an hour off each day for lunch. He finishes work at 4:30 P.M. About how many pieces of mail does Dan deliver each day?

   about _____S pieces of mail

Problem Solving: Skill
Find an Estimate or Exact Answer

Solve. Tell if you found an estimate or exact answer.

1. There are 4 bags of peanuts. Each bag contains 89 peanuts. Are there more or less than 400 peanuts in all?

   ____

   Did you find an estimate or exact answer? ______________

2. Each pack has 200 sheets of paper. How many sheets of paper are in 3 packs?

   ______ sheets

   Did you find an estimate or exact answer? ______________

Solve.

3. A set of markers cost $4. There are 30 art students in class. Each student will buy a set of markers. How much will the class spend on markers?

   ____

   Did you find an estimate or exact answer? ______________

4. A set of paintbrushes cost $8. The art teacher needs 18 sets. If she has $180 in her budget to buy supplies, does she have enough money for the paintbrushes?

   ____

   Did you find an estimate or exact answer? ______________

Solve.

5. Each of the 4 third-grade classes has 19 students. Each of the 3 fourth-grade classes has 29 students. About how many more students does fourth-grade have than third grade?

   about ______ students

   Did you find an estimate or exact answer? ______________

6. The school cafeteria ordered 510 hot dogs for the week for lunch. Each school day, 90 students ordered hot dogs. How many more hot dogs will they need to order for next week so they will have 510 hot dogs again?

   ______ hot dogs

   Did you find an estimate or exact answer? ______________

18.1 Explore Multiplying 2-Digit Numbers

Use models to solve.

1. Vin works 16 days each month. How many days does he work in 2 months?

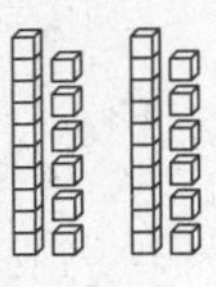

_______ days

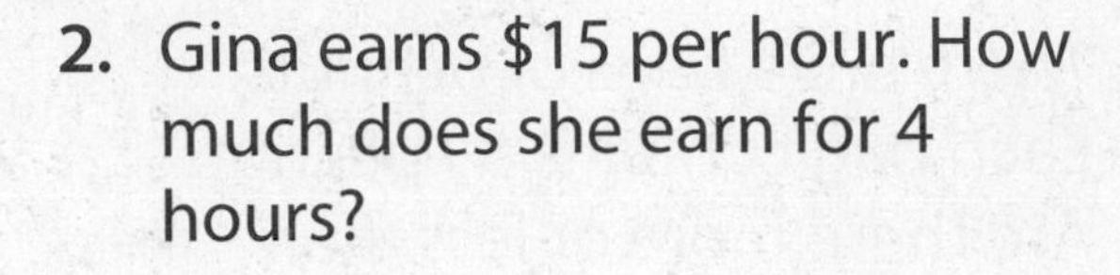

2. Gina earns $15 per hour. How much does she earn for 4 hours?

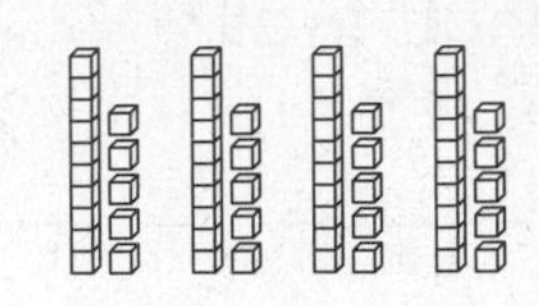

____

Use models to solve.

3. Each business phone call Mr. Harlow makes lasts for 26 minutes. He made 6 business calls on Monday. How many total minutes did the calls take?

_______ minutes

4. At the end of each month, Jackie saves $72. If she has done this for January through May, how much has she saved?

____

Solve. Use models if you need help.

5. A coyote can run about 43 miles per hour. A giraffe can run about 32 miles per hour. If both animals ran at these speeds, how far away would the giraffe be from the coyote at the end of 4 hours? Assume the animals started at the same place and ran in the same direction.

_______ miles away

6. In March, Mr. Roth had $546 in his checking account. He wrote checks for his electric and phone bills for March through June. Each month he pays $34 for his electric bill and $56 for his phone bill. How much was left in his checking account at the end of June?

____

Multiply 2-Digit Numbers

Solve.

1. Carol speaks to her grandmother on the phone 3 times a week for 25 minutes each time. For how many minutes each week does Carol speak to her grandmother?

   ________ minutes

2. Sanjay has 3 boxes of movies on DVD. Each box holds 18 DVDs. How many DVDs does he have in all?

   ________ DVDs

Solve.

3. Each lesson in a science book has 34 pages. There are 8 lessons in the book. If Ed reads the whole science book, how many pages will he have read?

   ________ pages

4. There are 25 white paper clips and 75 silver paper clips in each box. How many silver clips are in 9 boxes?

   ________ silver paper clips

Solve.

5. The school store sells folders at a price of $0.48 each. An office supply store has a special sale of 6 folders for $3. Which store sells 6 folders for less money? Explain.

   ____

6. A grocery store sells bags of peanuts for $0.45 each, bags of cashews for $0.78 each, and bags of almonds for $0.95 each. Bryan pays for 2 bags of each kind of nut with a $5 bill. How much change should he get back?

   $________

18.3 Problem Solving: Strategy
Make a Graph

Make a pictograph. Use the table to solve.

Votes for Favorite Color

Color	Votes
Red	33
Blue	66
Green	44

1. Make a pictograph to show the data in the table. Use dots to represent the number of votes. Each ● stands for 11 votes.

____

2. Which color received the most votes?

____

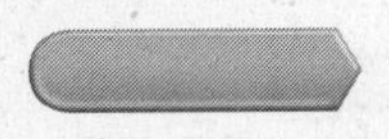

Make a pictograph. Use the table to solve.

Swim Club Age Groups

Ages	Number of Teammates
6–8	48
9–11	36
12–14	108
15 and up	132

3. Make a pictograph to show the data in the table. Each 🧍 stands for 12 teammates.

____

4. How many more teammates are in the oldest age group than in the youngest age group?

_____ more

Make a pictograph. Use the table to solve.

Place-Value Models Kits

Model	Number of Models
Ones	225
Tens	175
Hundreds	75
Thousands	25

5. Make a pictograph to show the data. What should each ● stand for: 25 models, 35 models, or 45 models?

_____ models

6. If you lose 100 Ones Models, how would your pictograph change?

____

Multiply Greater Numbers

Solve.

1. Pocket Electronics store has 2 floors of products. Each floor has 115 CD players. How many CD players are in the store?

   _______ CD players

2. Each rack in the electronics store has 161 DVDs. There are 3 racks. How many DVDs are in the store?

   _______ DVDs

Solve.

3. It takes 494 gallons of paint to paint the outside of the school building. The building is painted every year. How many gallons of paint are used after 5 years of painting the building?

   _______ gallons

4. Seven new classrooms are being added to the Lumberton Elementary School. The floor in each classroom takes 1,276 tiles. How many tiles are needed to cover all of the floors of the new classrooms?

   _______ tiles

Solve.

5. You must climb 1,060 steps to reach the second floor of the Eiffel Tower in Paris, France. Andre walked up and down three times. How many steps did he walk up and down all together?

   _______ steps

6. Neil is an airline pilot. On each of his first 4 trips, he flew 3,456 miles. On his last trip he flew 8,569 miles. How many miles did he fly in all 5 trips?

   _______ miles

Choose a Computation Method

Choose a computation method and solve.

1. Justin inflated 8 bunches of balloons. There are 10 balloons in each bunch. How many balloons did Justin inflate?

   _______ balloons

   Tell which method you used.

   ____

2. The school auditorium has 6 tables. Each table has 13 science projects. How many science projects are there in all?

   _______ science projects

   Tell which method you used.

   ____

Choose a computation method and solve.

3. There are 6 theaters at the Chelsea Cinema. Each theater holds 229 people. What is the total number of people that the Cinema can hold?

   _______ people

4. In one hour, the Delmont Bakery makes 2,000 doughnuts. How many doughnuts do they make in 8 hours?

   _______ doughnuts

Solve.

5. The owner of the Fluffy Feathers Pet Store is ordering two kinds of parrots. Grey Parrots cost $1,521 each, and Cape Parrots cost $950 each. The store is buying 3 Grey Parrots and 4 Cape Parrots. How much will the store owner spend on the parrots? _______

6. The pet store owner needs to order birdseed. The owner can order birdseed in 20-pound bags for $23.51 each, or in 25-pound bags for $25.03 each. The pet store sells an average of 200 pounds of birdseed each month. How much would the owner pay for 200 pounds of birdseed in 20-pound bags? _______ How much would the owner pay for 200 pounds of birdseed in 25-pound bags? _______ Should the owner buy the 20-pound bags or the 25-pound bags? Why? ______________________

Explore Dividing Multiples of 10

Use the models to solve.

1. After working for 3 weeks, Pat earned $60.

   How much did he earn each week?

   _____ each week

2. The office supply store has 2 boxes of folders left that are on sale. There are 80 folders in all.

   How many folders are in each box?

   _______ folders

Solve. Use models if you need help.

3. The computer printer has 240 sheets of paper in it. Each student prints out an 8-page book report. Now the printer is empty. How many students printed out their reports?

   _______ students

4. Mr. Wilson will give out 120 textbooks to the class. Each student will get 6 textbooks. How many students are in the class?

   _______ students

Solve. Use models if you need help.

5. Theo spent a total of $560 in 8 weeks. He spent the same amount each week. He spent $30 per week on food, and he paid bills with the rest of the money. How much did he spend each week on bills?

   _____ each week

6. It took 490 seconds for Megan to download 2 data files for science and 5 data files for math. Each file took the same amount of time to download. Was this more than or less than 1 minute for each file? Explain.

   ____

   ____

Algebra: Division Patterns

Use a pattern to divide.

1. A father divides 40 cookies equally among his 4 children. How many cookies does each child get?

   ________ cookies

2. Dave packs 60 books into 3 boxes. Each box has the same number of books. How many books are there in one box?

   ________ books

Divide. Use patterns.

3. There are 320 chairs in the auditorium. If there are 8 equal rows of chairs, how many chairs are in each row?

   ________ chairs

4. Speakers at a town meeting talk for 6 minutes each. The speakers at last night's meeting talked for a total of 120 minutes. How many speakers talked at the meeting?

   ________ speakers

Solve.

5. Sandra spent $476 on office supplies. She bought a fax machine for $130, a conference table for $96, and 5 chairs. If each chair was the same price, how much did one chair cost?

   ____

6. Bianca drove 257 miles on Monday and 193 miles on Tuesday. If she drove a total of 9 hours over both days, how many miles per hour did she drive?

   ________ miles per hour

Estimate Quotients

Estimate to solve.

1. Amy feeds the 4 class hamsters the same amount of food each day. She has 22 ounces of food. About how many ounces of food does each hamster get per day?

   about ______ ounces

2. In art class, Cory is making paper chains. It takes him 6 minutes to make each chain. There are about 28 minutes left in class. About how many more chains can he make?

   about ______ more chains

Estimate to solve.

3. Lorrie is emptying her sister's wading pool with a pump. The pool holds 142 gallons. Each minute the pump removes 7 gallons of water. About how many minutes will it take to empty the pool?

   about ______ minutes

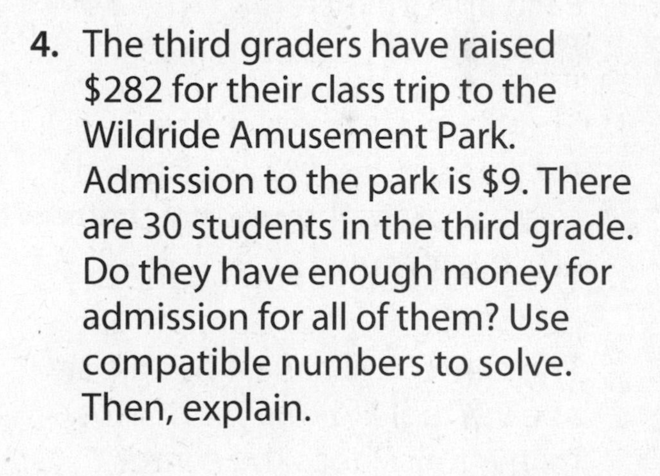

4. The third graders have raised $282 for their class trip to the Wildride Amusement Park. Admission to the park is $9. There are 30 students in the third grade. Do they have enough money for admission for all of them? Use compatible numbers to solve. Then, explain.

   ____

   ____

Solve.

5. Nina saved $3.50 each week. After 7 weeks she spent half of her savings on a new shirt, and kept the other half in the bank. About how much did the shirt cost?

   about ______

6. The total distance around the Kennington Village Square is 3,928 feet. About how long is one side of the village square?

   about ______ feet

19.4 Explore Division

Use the models to solve.

1. An apartment building has 48 windows. There are 4 apartments in the building. Each apartment has the same number of windows.

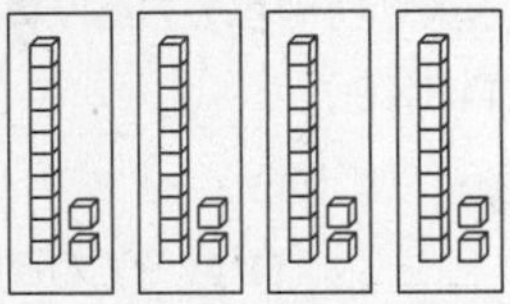

   How many windows are there in each apartment?

   _______ windows

2. An office building has 33 trash cans. They are divided equally on 3 floors of the building.

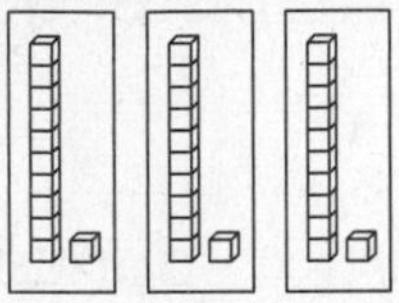

   How many trash cans are there on each floor?

   _______ trash cans

Solve. Use models if you need help.

3. The Recreation Center had 72 people sign up for Game Night. The center will have 4 people sit at each table for the night. How many tables does the center need to set up?

   _______ tables

4. Kenya is mailing 6 jars of her special sauce to her cousin. The total weight of all 6 jars of sauce is 96 ounces. How many ounces does each jar weigh?

   _______ ounces each

Solve. Use models if you need help.

5. A donut shop just baked 84 donuts. They charge $4 for each box of 6 donuts. How much will they make if all of the donuts are sold in boxes of 6?

   ____

6. Alan has 81 model cars. He will keep 3 cars on display on a shelf in his room. He will store the remaining cars in boxes with 6 cars to a box. How many boxes does he need to store the cars?

   _______ boxes

Divide 2-Digit Numbers

Estimate first. Then divide.

1. At the Royce School there are 48 cars in the teachers' parking lot. The same number of cars are parked in each of 3 rows. How many cars are parked in each row?

   _______ cars

2. The art teacher has 56 paintbrushes. He puts the same number of brushes into 4 different sections of his art box. How many brushes are in each section?

   _______ brushes

Solve.

3. Clare works at a laundromat. She will wash 72 pairs of pants. The washing machine can wash 6 pairs of pants for each load of laundry. How many loads of laundry will she need to do to wash all of the pants?

   _______ loads

4. For a class project, Marty has 69 pieces of pasta. He is pasting the pasta into 6 equal rows on poster board. How many pieces will be in each row?

   _______ pieces

   How many will be left over?

   _______ pieces

Solve.

5. Gina has 42 pennies in her bank and 26 pennies in her wallet. She wants to exchange the pennies for nickels. How many nickels will she get?

   _______ nickels

   How many pennies will she still have left?

   _______ pennies

6. There are 18 boys and 17 girls in the third grade. Each day 3 students will give an oral report. How many days will students be giving oral reports?

   _______ days

   How many students will give their report on the last day?

   _______ students

Problem Solving: Skill
Interpret the Remainder

Solve.

1. The 17 students in the science club are going to the zoo. Club parents will drive them. If they can fit up to 5 students in each car, how many cars will they need?

   _______ cars

2. The teacher gathers 26 crayons and puts them back in their boxes. Each box holds 8 crayons. Any left over crayons are put in a drawer. How many crayons are put into the drawer?

   _______ crayons

Solve.

3. Each guest at a birthday party will get one box of juice. There are 58 guests at the party. How many packs of juice are needed if there are 6 boxes of juice in a pack?

   _______ packs

4. Hal makes balloon animals. He uses 5 balloons to make a balloon giraffe. If a package has 96 balloons, how many balloon giraffes can he make?

   _______ balloon giraffes

Solve.

5. Mia is allowed to talk on the phone to her friends from 3:15 P.M. to 4:15 P.M. She decides to call 7 friends and talk to each one for the same number of minutes. What is the greatest number of whole minutes she can talk to each of her friends?

   _______ minutes

6. A computer store sells blank disks in packages of 8 for $6. Individual disks cost $1.25 each. What is the least amount you can spend if you want to buy 60 blank disks?

   ____

Divide 3-Digit Numbers

20.1

Solve.

1. Mother divides 147 carrots equally into plastic snack bags. If she puts 7 carrots in each bag, how many plastic snack bags does she need?

   _______ snack bags

2. Mrs. Ruiz has $1.26. She gives an equal amount to each of her 3 children. How much does each child get?

   $_______ each

Solve.

3. Elizabeth is making large candles. She has 228 pounds of wax. Each candle will use 6 pounds of wax. How many candles can Elizabeth make?

   _______ candles

4. Myla is the coach of the Pearson Sack Race Club. The club has 156 sacks and 7 members. After practice, each member takes the same number of sacks home. Myla takes home any sacks that are left over. If each member takes home as many sacks as possible, how many does Myla take home?

   _______ sacks

Solve.

5. Larry plays on the school basketball team. Larry scored a total of 134 points in 5 games. He scored the same number of points in each of the first 4 games. In the last game he scored 30 points. How many more points did he score in the last game than in any of the other 4 games?

   _______ more points

6. Rick plans to make 6 large birdhouses and 2 small ones. He will use a total of 148 nails. Each small birdhouse uses 8 nails. If each large birdhouse uses the same number of nails, how many nails will be used for each?

   ____

Quotients with Zeros

Solve.

1. Alice has 202 golf balls. She is putting the same number of balls into 2 pails. How many will she put into each pail?

   _______ golf balls

2. The library will store 424 magazines. The same number of magazines will be stored in each of 4 boxes. How many magazines will be stored in each box?

   _______ magazines

Solve.

3. The Thomason Gallery has 621 photographs. The same number of photographs are displayed in each of 3 rooms. How many photographs are shown in each room?

   _______ photos

4. For graduation Erica received $412. She wants to divide the money equally among 4 charities. What is the most money she can give to each charity?

   $_______ to each charity

Solve.

5. Last year Heather jogged at 6 different parks. She ran a total of 617 miles. Could she have run the same number of miles at each of the parks? Explain.

   ____

6. Nine planes flew out of Hampton Airfield. They flew a total of 1,903 miles. One of the planes flew 263 miles. The remaining planes flew equal distances. How many miles did each of the remaining planes fly?

   ____

Choose a Computation Method

Solve. Tell which method you use.

1. A helicopter flew at the same speed for 2 hours. It flew 200 miles. How many miles did it fly in 1 hour?

   ________ miles

   Which method did you use?

   ____

2. At Brennan's Hobby Shop there are 42 airplane models. The same number of models are displayed in each of 3 glass cases. How many airplane models are in each case?

   ________ models

   Which method did you use?

   ____

Solve. Tell which method you use.

3. The Rumson family made a giant, 129-inch snowman. Each day the snowman melted 8 inches. How many days did it take for the snowman to melt completely?

   ________ days

   Which method did you use?

   ____

4. Yesterday the three hippopotamuses at the zoo ate 390 kilograms of grass. Each hippo ate the same amount of grass. How much grass did each hippo eat?

   ________ kilograms

   Which method did you use?

   ____

Solve. Tell which method you use.

5. Each day this spring, an average of 9 songbirds arrived from the south in Suttonville. There are now 567 songbirds in town. How many weeks did it take for all of them to arrive?

   ________ weeks

6. Mr. Walsh drives a total distance of 200 miles to and from work each week. He works 5 days per week. How many miles does Mr. Walsh drive to work?

   ________ miles

   Which method did you use?

   ____

20.4 Problem Solving: Strategy Choose a Strategy

Solve. Tell which strategy you chose.

1. Every 4th dancer in a chorus line wears a hat. There are 30 dancers in the line. How many dancers wear hats?

   _______ dancers

   Which strategy did you choose?

   ____

2. There are 2 shelves in a closet. Each shelf has 3 sweaters. Each sweater has 10 buttons. How many buttons in all?

   _______ buttons

   Which strategy did you choose?

   ____

Solve. Tell which strategy you chose.

3. Jeff, Gordon, and Vince buy 2 slices of pizza each for lunch. In all, they spend $18 for lunch. What is the cost of one slice of pizza?

   $______

   Which strategy did you choose?

   ____

4. Construction paper comes in 6 colors, and 90 sheets to a pack. There are the same number of sheets for each color in the pack. How many sheets are there of each color? _______ sheets

   Which strategy did you choose?

   ____

Solve. Tell which strategy you chose.

5. It took Ben 65 minutes to finish his homework by 4:30 P.M. He had homework in 5 different subjects and worked on each for the same number of minutes. If he did math first, what time did he finish math?

   ____

   Which strategy did you choose?

   ____

6. There are 8 coins on the table. Their total value is $0.55. What coins are on the table?

   ____

   How many of each?

   ____

   Which strategy did you choose?

   ____

Explore Lengths

Solve.

1. Max drew a line that was 3 counting cubes long. Draw the line that Max drew.

____

2. How many paper clip units long is the carrot?

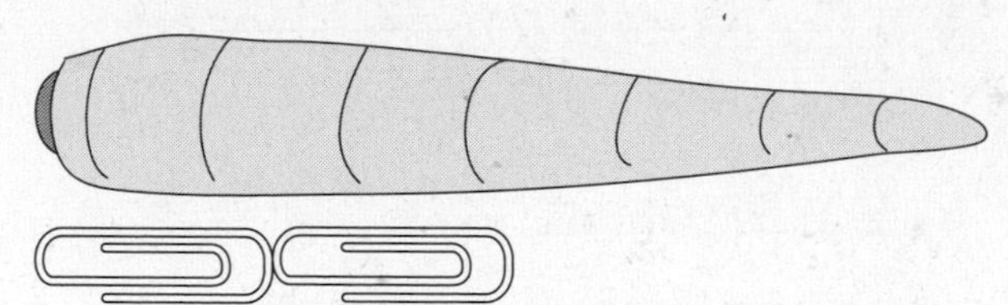

____

Solve.

3. Harry uses connecting cubes to measure a comb. It measures 6 connecting cubes long. Will it be more or less than 6 inches?

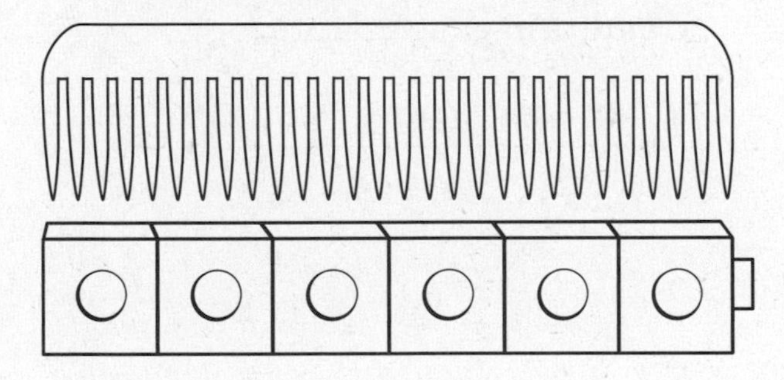

____

4. Frank measures the width of his desk in paper clip units. His desk measures 28 paper clip units. How can he use this information to find the width of his desk in inches?

____

Solve.

5. Richard and Tanya measured the length of a long piece of chalk. Richard said the chalk was 8 units long, and Tanya said it was 3 units long. Explain why both measurements might be correct.

____

6. Delon measures the length of his shoe in paper clips and in inches. His shoe measured 8 inches, and 5 paper clips. Are the paper clips longer or shorter than an inch?

____

Explore Customary Units of Length

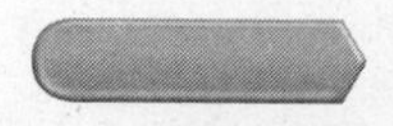

Solve.

1. How many inches long is the pencil shown in the picture?

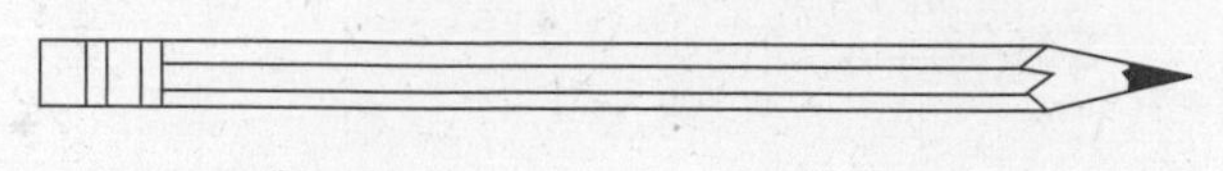

____

2. What measuring tool would you use to measure the length of your foot?

____

Solve.

3. Kira went on a hiking trip. She hiked for 4 hours in one day. What unit of measurement would you use to tell the distance she walked in 4 hours?

____

4. Hector wants to measure the length of his dog's tail. What tool is he most likely to use?

____

What unit of measure?

____

Solve.

5. Carrie wants to make a paper cover for her math book. Explain how you would measure a book to find out how much paper to use.

____

6. There will be a new floor in the gym. The principal had to find the length and width of the gym. The measurements were written on the order form, but the units of measure were accidentally erased. The order form said the gym floor is 15_____ by 20 _____. What units of measure were most likely erased?

____

Customary Units of Capacity

Solve.

1. Alan has a glass of juice with lunch. Did he drink about 2 cups or 2 gallons of juice?

____

2. A container of milk holds 4 quarts. Is that more, less, or the same as 1 gallon?

____

Solve.

3. Krista's teacher asked her how much water she thinks a large bathtub can hold. Krista said, "About 48." Krista did not say the unit of measure. What unit should she have said?

____

4. A punch bowl at a party has more than 1 gallon of juice in it. Guests take juice from the bowl using cups. What is the fewest number of cups of juice in the punch bowl?

____

Solve.

5. Angie wants to fill a gallon fish bowl with water. She will fill a drinking glass with tap water, then pour it into the bowl. About how many times will she probably need to fill the glass in order to fill the fish tank? Explain.

____

____

6. Mrs. Polk made a gallon of soup for her children for lunch. The 3 children ate 2 small bowls of soup each. A bowl is a little more than 1 cup. About how many quarts of soup do you think is left? Explain.

____

____

Customary Units of Weight

Solve.

1. Jake bought a bag of apples. Does it weigh about 3 pounds or 3 ounces?

____

2. A loaf of bread weighs about 1 pound. About how many ounces does it weigh?

____

Solve.

3. Peggy told her mom that the new puppy weighs, "about 3." She didn't give the unit. What unit of weight should she have said after the number 3?

____

4. A customer weighs an orange on the scale in the fruit store. The orange weighs 5 ounces. The customer does not want to buy more than 1 pound of oranges. What is the greatest number of oranges the customer can buy? Explain.

____

Solve.

5. Bart is putting groceries in a paper bag. He bought a loaf of bread, a cupcake, a bag of 10 apples, a bag of flour, and a bag of chips. He puts 3 items in one bag. Now the bag weighs about 2 pounds. Which 3 items did he probably put in the bag?

____

6. A bag of white potatoes weighs 2 pounds. A bag of sweet potatoes weighs 16 ounces. Tell how you could rearrange the potatoes so that you have two bags of potatoes that weigh the same.

____

Convert Customary Units

Solve.

1. The bedroom window is 2 feet wide. How many inches wide is the window?

____

2. A large container of milk contains 8 quarts. How many gallons does it hold?

____

Solve.

3. Esther drinks 8 glasses of water every day. Each glass measures 2 cups. How many quarts of water does Esther drink a day?

____

4. A small watermelon weighs about 5 pounds. How many ounces does the watermelon weigh?

____

Solve.

5. Mark's smaller cat weighs 8 pounds. His larger cat weighs 14 ounces more than the smaller one. How many ounces does Mark's larger cat weigh?

____

6. Megan needs 6 feet of ribbon to finish a sewing project. One roll of ribbon has 22 inches left and another has 1 yard and 3 inches left. How much more ribbon will Megan need to buy to finish her sewing project?

____

Problem Solving: Skill Check for Reasonableness

Solve.

1. Ruben is 4 ft tall. His sister is 50 inches tall. Ruben says he is shorter than his sister. Is he correct?

____

2. Peter bought 3 quarts of milk and 1 gallon of juice. He said that he bought more milk than juice. Is he correct?

____

Solve.

3. Each of Paul's guinea pigs weighs 10 ounces. Maureen's guinea pig weighs a pound. She says her pet weighs more than Paul's two pets combined. Is she correct? Why or why not?

____

4. The living room is 12 ft long and 15 ft wide. The bedroom is 4 yards long and 5 yards wide. Jan said the living room and the bedroom are the same size. Is she correct? Why or why not?

____

Solve.

5. A fish store sells lobster for $2 per ounce. Harris buys a lobster that is 1 pound 3 ounces. He pays for it with two $20 bills. Harris says he should get back $5 change. Is Harris correct? Why or why not?

____

____

6. The 3 children in the Rodriguez family are the following heights: Pablo is 4 ft 6 in. tall; Maria is 58 in. tall; and Juan is 1 yard 2 ft tall. Mother says her tallest child is 6 inches taller than her shortest child. Is she correct? Why or why not?

____

Explore Metric Units of Length

22.1

Solve.

1. Which is about 1 centimeter long: a fingernail or a pencil?

____

2. Which metric unit would you use to measure the distance between the front of the classroom and the back of the classroom?

____

Solve.

3. A tomato plant is a little less than 1 yard in height. How can you describe this length in metric units?

____

4. Estimate the length of the crayon in centimeters.

about ________ centimeters

red

Now use a centimeter ruler to measure the length of the crayon to the nearest centimeter.

________ centimeters

Solve.

5. Gail and Eric estimated the length of a screwdriver. Eric estimated about 11 cm. Gail said about 10 cm. They used a ruler to measure it, and found it was exactly 10.8 cm. Who had a better estimate? Why?

____

6. Estimate the length of your arm span to the nearest decimeter.

About ________ decimeter

Now use a metric ruler to check your estimate. What is the length of your arm span?

________ decimeters

Was your estimate close? ______

Explain.

____

22.2 Metric Units of Capacity

Solve.

1. Tanya has a cold. What metric unit of measure should she use to measure the amount of liquid cold medicine she should take?

____

2. Circle the best estimate for each.

Bottle top:	40 L	40 mL
Bucket:	7 L	7 mL
Mug:	480 L	480 mL

List the containers in order from least to greatest capacity.

____

Solve.

3. Lisa bought a liter of water. She drank 250 milliliters. How much water is left in the bottle?

____

4. The chef put 2 teaspoons of vanilla extract into a cake recipe. About how much vanilla extract is this, 2 mL or 2 L?

____

Solve.

5. Maurice used a glass that held 250 milliliters of liquid to fill a 2-liter bottle. He had to refill the glass several times to fill the bottle. How many times did he have to refill and pour the water from the glass to fill the bottle? Explain.

____

6. Below is Dena's recipe for fruit punch. Write the missing metric units of capacity for each ingredient.

Dena's Fruit Punch
Frozen Orange juice: 2 _______
Pineapple juice: 360 _______
Apple juice: 1 _______

____

Metric Units of Mass

22.3

Solve.

1. List the items below in order from least to greatest mass.

   sheet of paper   box of crayons
   feather

   ____

2. What metric unit of mass would you use to measure the mass of a pencil?

   ____

Solve.

3. A paper clip has a mass of 1 gram. A box of paper clips has a mass of 1 kilogram. How many paper clips are in the box?

   ____

4. A liter of water has a mass of about 1 kilogram. Matt carries a liter bottle of water and 750 grams of granola bars in his lunch box. Does the water or box of granola bars have a greater mass? Explain.

   ____

Solve.

5. Wendy estimates that 5 of her erasers have the same mass as her box of pencils. Each of her erasers has a mass of 150 grams. The box of pencils has a mass of 1 kilogram. Is her estimate correct? Explain.

   ____

6. One side of a balance scale has a textbook with a mass of 2 kilograms. How many paper clips would you need to put on the other side of the scale to balance it if each paper clip has a mass of 1 gram?

   ____

22.4

Convert Metric Units

Solve.

1. A bedroom door is 3 meters tall. What is its height in centimeters?

____

2. A kitten has a mass of 2,000 grams. What is its mass in kilograms?

____

Solve.

3. Each paper cup holds 50 milliliters of water. How many cups of water would you need to fill a 2-liter bottle?

____

4. An apple has a mass of 200 grams. What is the mass of 10 apples in kilograms?

____

Solve.

5. The art teacher displays several pictures side by side across a wall. Each picture is about 24 cm wide. The line of pictures measures 1 meter from end to end. How many pictures are displayed in the line? Explain.

____

6. Greg has 6,500 milliliters of water in a large jug. He wants to pour the water into liter bottles so that the water will be easier to store. What is the smallest number of bottles he will need? Explain.

____

Problem Solving: Strategy Logical Reasoning

Solve.

1. Jan is 3 inches taller than Dan. Ellie is 2 inches taller than Jan. If Ellie is 54 inches tall, how tall are Jan and Dan?

____

2. One bag of rice weighs 16 ounces and another bag weighs 8 ounces. How much rice can you pour from the larger bag into the smaller bag so that both weigh the same?

____

Solve.

3. Rod is less than 17 years old. The sum of the digits in his age is even and greater than 4, but both digits are odd. How old is Rod?

____

4. Copper wire comes in rolls of 2 ft, 1 yard, or 60 inches. An electrician bought 5 rolls that totaled exactly 216 inches of copper wire. How many of each roll did he buy?

____

Solve.

5. Al, Beth, Chuck, and Dean are standing in line. Chuck is standing in back of Dean. Beth is standing in front of Al. Dean is not first. List the names in order from first to last in line.

____

6. Mr. Moses needs 13 liters of water in a tank. He has a 7-liter bottle and a 10-liter bottle. How can he use the bottles to pour exactly 13 liters of water into the tank?

____

Temperature

Solve.

1. The temperature outside is 84°F. Should you wear a coat?

2. It feels cold outside today. Is the temperature probably 25°F or 25°C?

Solve.

3. At 6:00 A.M. the outdoor temperature is 15°F. If the temperature goes up 10°F in the next hour, what is the temperature at 7:00 A.M.?

4. Grace invites her cousin Darla to visit her in March. Grace tells Darla the temperature usually ranges from 35°C to 39°C. What kind of clothing should Darla pack for her visit?

Solve.

5. The outdoor temperature at 8:00 P.M. was 24°C. The temperature dropped an average of 3 degrees per hour until 1:00 A.M. What was the temperature at midnight?

6. The oven temperature rises at about 25°F per minute. The recipe says to preheat the oven to 350°F. Jane turns on the oven and it starts at 100°F. In how long will it reach 350°F?

3-Dimensional Figures

Solve.

1. Penny had a drink in a container shaped like a rectangular prism. What did Penny drink?

   ____

2. What is the shape of the orange juice container?

   ____

Solve.

3. Peter folded the paper net shown below along the lines to make a 3-dimensional figure. What figure did he make?

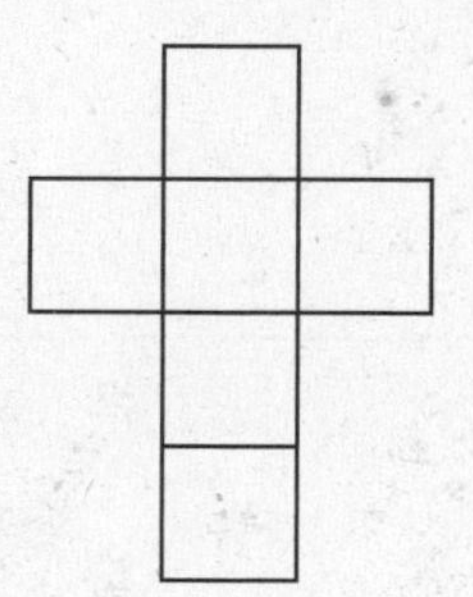

   ____

4. Ricky traced around the bottom of a box shaped like a pyramid. What shape did Ricky draw?

   ____

Solve.

5. Which of these pencil parts is shaped like a cylinder? A cone?

   ____

6. Justin made a 3-dimensional figure out of a net that had 2 equal-sized circles and a rectangle. What figure did he make?

   ____

2-Dimensional Figures

 Solve.

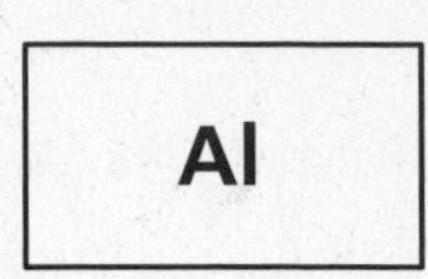

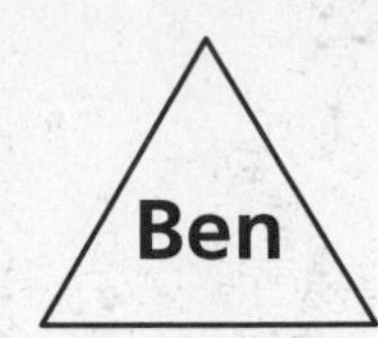

1. What shape is Cora's nameplate?

____

2. One of the students' nameplates is in the shape of a rectangle. Whose nameplate is it?

____

 Solve.

3. The window in Nan's bedroom is made up of 4 rectangular panes of glass. How many angles in all does the window have?

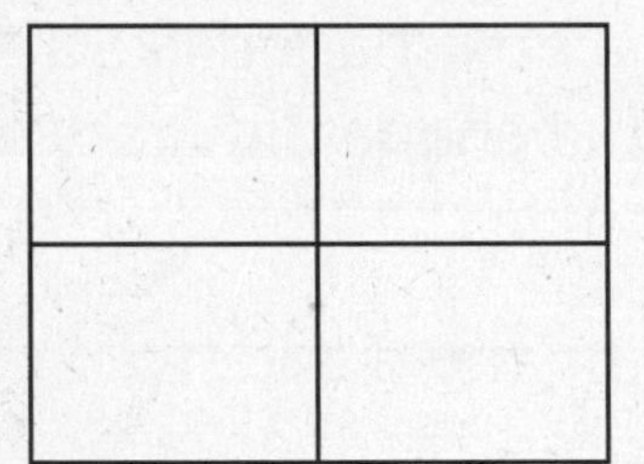

____

4. The "Welcome" sign on Cally's front door has 3 straight sides and 3 angles. What shape is Cally's welcome sign?

____

 Solve.

5. Louis says that a rectangle can be a square, but it can not be a triangle. Is Louis correct? Why?

____

6. Can you form a rectangle from the two triangles shown below? Draw a picture to show why or why not.

____

Lines, Line Segments, Rays, and Angles

23.3

Solve.

1. Evan drew a square. How many right angles did he draw?

____

2. What kind of lines are the rungs on the ladder examples of?

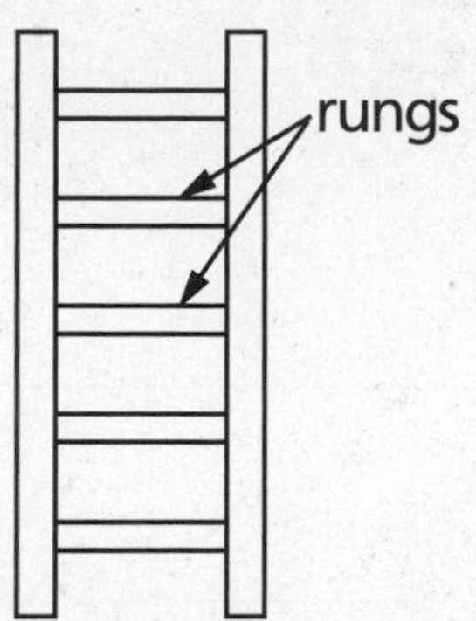

____

Solve.

3. The door to Ellen's bedroom has two panels on it. One is in the shape of a rectangle and the other is in the shape of a triangle. How many line segments do the panels have in all?

____

4. For one question on a math test, Tanika had to draw a triangle with one angle greater than a right angle and two angles less than a right angle. Draw a picture to show Tanika's triangle.

____

Solve.

5. Yasmin drew a figure that is part of a line. Her figure had one endpoint and went without end in the other direction. What figure did Yasmin draw?

____

6. The math teacher says the lines below are intersecting even though they do not meet. Explain why this is true.

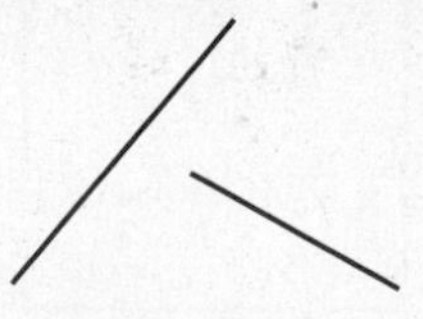

____

23.4

Polygons

Solve.

1. Each tile on a floor has 5 sides and 5 angles. What shape is each of the tiles?

____

2. What is the shape of the stop sign?

____

Solve.

3. Peter made a hexagon using 6 toothpicks. He now wants to change the hexagon into an octagon. How many more toothpicks does he need?

____

4. Is a circle a polygon? Why or why not?

____

Solve.

5. Four students were asked to name the figure below. Each student answered differently, but each was correct. What were the students' answers?

____

6. Lana drew a design using the same number of hexagons and octagons. The design has a total of 42 sides. How many hexagons are in the design?

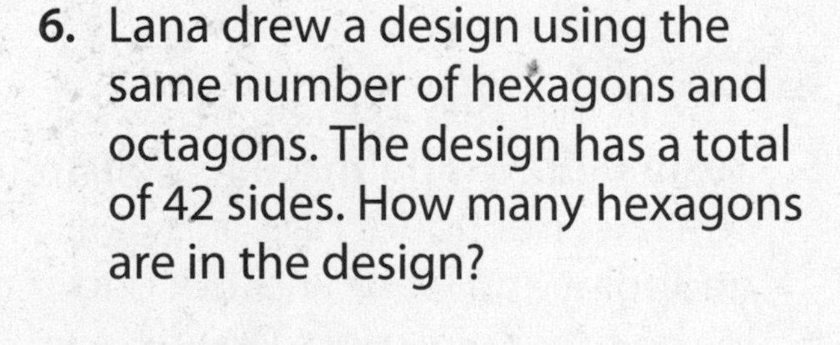

_______ hexagons

Triangles

Solve.

1. Elaine drew a triangle with only two of the sides the same length. What kind of triangle did she draw?

____

2. The angles of a window shaped like a triangle are each less than a right angle. What kind of triangle is the window shaped like?

____

Solve.

3. An equilateral triangle has 2 sides that are each 4 inches long. What is the length of the third side? How do you know?

____

4. Billy has a pattern block of a right triangle and Laura has one of an obtuse triangle. Tell how the pattern blocks are alike.

____

Tell how they are different.

____

____

Solve.

5. The flower bed in the park by Amy's house is in the shape of an equilateral triangle. One side of the triangular flower bed is 40 feet long. Amy walked around all three sides of the triangle. How far did she walk?

____

6. Ron is building a structure in the shape of a rectangular pyramid. He will use 5 sheets of plywood to build the pyramid. How many pieces of triangular plywood will he use?

_______ pieces

Explain.

____

23.6 Quadrilaterals

Solve.

1. Rhonda makes two different quadrilaterals with toothpicks. Both have sides of the same length. What two quadrilaterals did she make?

____

2. The pattern blocks in a box are quadrilaterals except for one. What shape could the block that is not a quadrilateral be?

____

Solve.

3. Three picture frames are on a dresser. Two are shaped like squares and the other is shaped like a trapezoid. How many sides are there in all in the frames?

_________ sides

4. Collin says that all quadrilaterals are polygons, but not all polygons are quadrilaterals. Is he correct? Explain.

____

Solve.

5. Four students were asked to draw a parallelogram. Each drew a different shape, but each was correct. Explain how that can be.

____

6. Are both of the shapes shown below trapezoids? Explain.

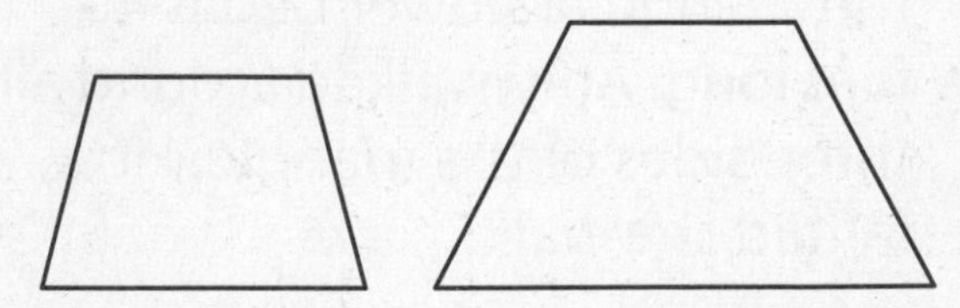

____

Problem Solving: Skill
Use a Diagram

Use the diagram to solve.

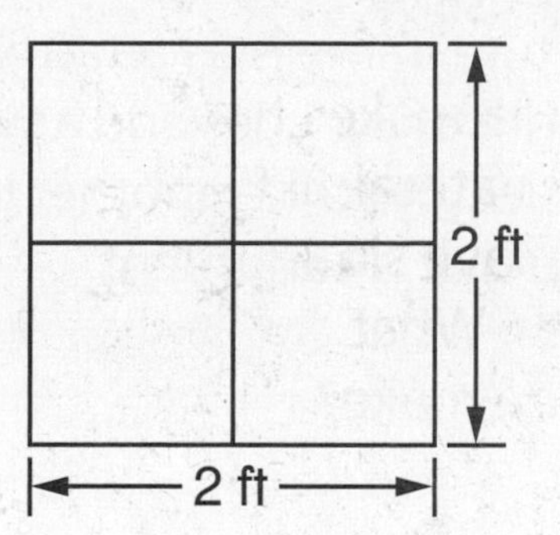

1. The diagram shows a window. What are the missing labels for the other 2 sides of the window?

____

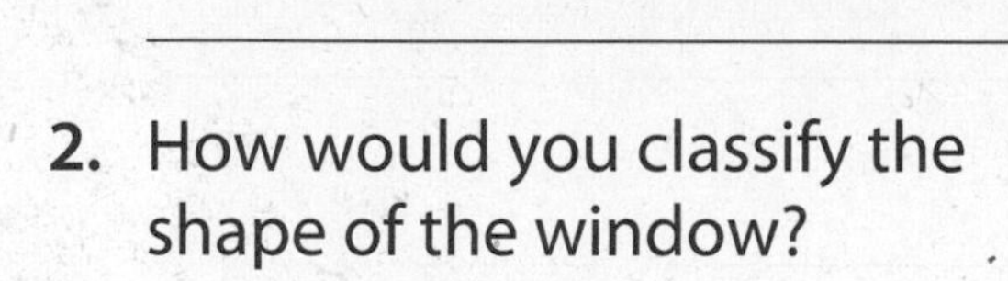

2. How would you classify the shape of the window?

____

Use the diagram to solve.

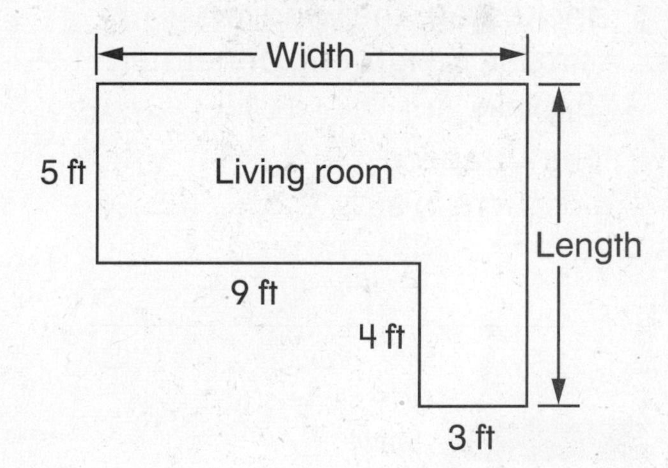

3. What is the width of the living room? Show your work.

____

4. What is the length of the living room? Show your work.

____

Use the diagram to solve.

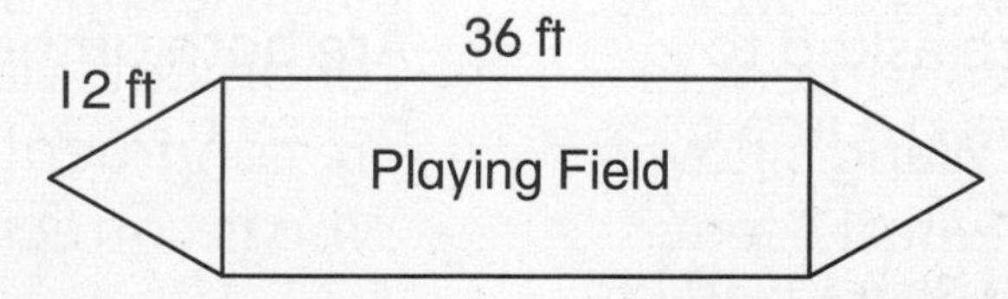

5. The playing field is made up of a rectangle with equilateral triangles at each end. What are the missing labels on the diagram?

____

6. The Park Department will build a fence around the entire playing field. How many feet of fence are needed?

____

Congruent and Similar Figures

Solve.

1. A picture frame is a rectangle with a length of 8 inches and a width of 6 inches. Describe another picture frame that is congruent.

2. The two lenses in Ira's eyeglasses are circular in shape. They are the same size. Are they similar or congruent?

Solve.

3. The windows of a garage door are shaped like squares. The length of each window is 10 inches. Are the windows similar, congruent, or neither? Explain.

4. Tricia has 2 pattern blocks. One is an equilateral triangle. The other is a scalene triangle whose sides are of different lengths. Are Tricia's pattern blocks similar, congruent, or neither? Explain.

Solve.

5. Both Mrs. Wang's and Mr. Altero's flower beds are in the shape of a rhombus. Mr. Altero's flower bed is similar but not congruent to Mrs. Wang's. Mrs. Wang's flower bed has sides that are 12 feet long. What could be the lengths of the sides of Mr. Altero's flower bed? Explain.

6. Melissa says that all circles are congruent and similar. Hallie says all circles are similar, but not congruent. Mike says all circles are congruent, but not similar. Who is correct? Explain.

Explore Translations, Reflections, and Rotations

Solve.

1. The hearts on this side of the lampshade were traced from the same heart shape. How was the heart shape moved to make the other hearts?

____

2. How many times was the heart shape moved after making the first tracing?

____

Solve.

3. If you make a reflection of this arrow and flip it over the dotted line one time, how will it change?

____

4. How many times will you need to make a rotation of the arrow around a point for it to be pointing down?

____

Solve.

5. Describe the kind of movement you are performing when you move the hands of a clock.

____

6. What kind of movement are you performing when you open a window that goes up and down?

____

Explore Symmetry

Solve.

1. Of the 4 charms hanging on the bracelet, which has only 1 line of symmetry?

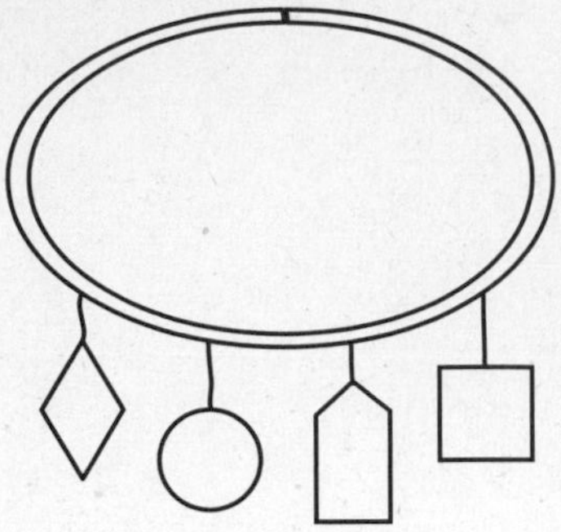

____

2. Which charm has the most lines of symmetry?

____

Solve.

3. Are the lines going through the banner lines of symmetry?

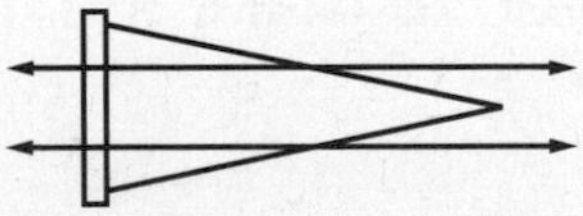

____

4. Does the rectangle have any other lines of symmetry in addition to the one shown? Explain.

____

Solve.

5. Damien cut out the isosceles triangle and square shown here. How many lines of symmetry do both shapes have all together?

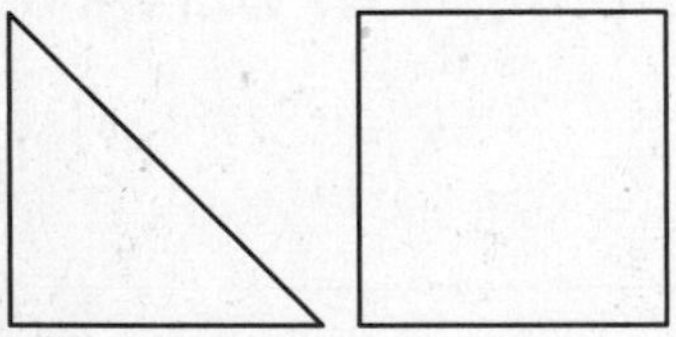

_______ lines of symmetry

6. Damien combined the shapes into a square with a triangle on top. How many lines of symmetry does the combined shape have?

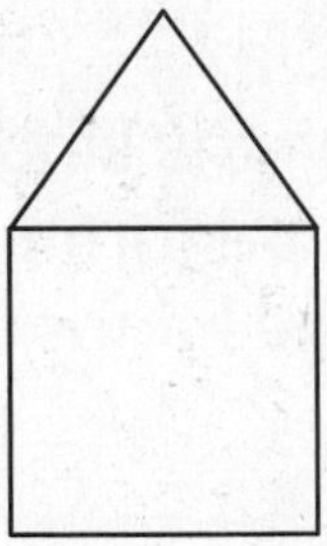

_______ line(s) of symmetry

Problem Solving: Strategy Find a Pattern

Find a pattern to solve.

△□□○○○△□□○○○△□□

1. Lisa is drawing a picture with the pattern of shapes shown above. What shape should she draw next to follow the pattern?

____

2. How can you use numbers to describe the pattern?

____

Find a pattern to solve.

3. Andy painted 30 small shapes along the wall behind his bed. The shapes were repeated in a pattern of 1 triangle, 3 squares, and 2 circles. How many circles did Andy paint on his wall?

_______ circles

4. Sally's flower garden has 10 rows of flowers. There are 4 flowers in the first row, 8 in the second, and 12 in the third. How many flowers are in the last row?

_______ flowers

Find a pattern to solve.

△△◁◁◁◁▽▽▽▽▽▽

5. Jenna made 3 designs using triangles. Describe the pattern of change from one design to the next.

____

6. If Jenna continues the design pattern, how many triangles will the seventh design have?

_______ triangles

Perimeter

Solve.

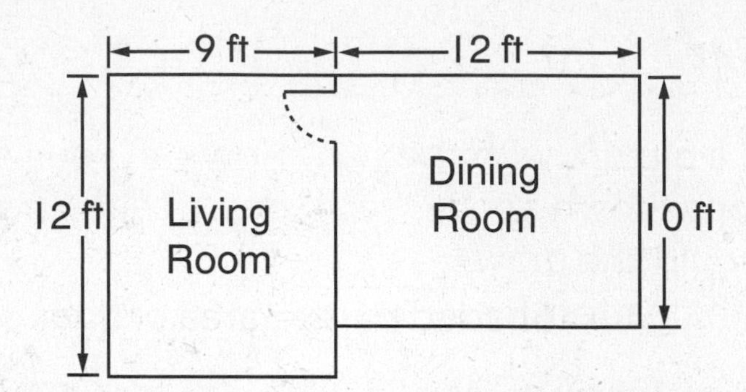

1. Jean will put a paper border around the 4 walls in the living room. How many feet of border paper will she use?

   ______ ft

2. How much border paper is needed to go around the walls of the dining room?

   ______ ft

Solve.

3. A rose garden is hexagonal in shape. Two of the sides are 14 feet and the remaining four sides are 16 feet each. How much fencing is needed to completely enclose the garden?

   ______ ft

4. Each of two bulletin boards is a 6 ft by 8 ft rectangle. How much border paper is needed to go around both boards?

   ______ ft

Solve.

5. The perimeter of the small gym is 120 feet. If the gym is rectangular in shape and it is 20 feet wide, how long is the gym?

   ______ ft

6. Zoe will sew lace around a rectangular tablecloth. The tablecloth is 60 inches long and 48 inches wide. If lace costs $2 per foot, how much will Zoe pay for the lace she needs?

   ______

Area

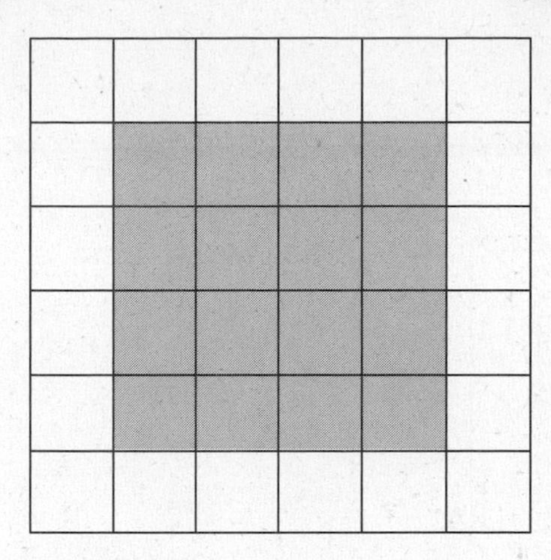

Shaded part = area of floor covered by a rug

 **Use the model above to solve.**

1. What is the area of the rug in square units?

   ________ square units

2. What is the area of the whole floor, including the rug covered area?

   ________ square units

 **Use the model above to solve.**

3. What is the area of the floor that is not covered by a rug?

   ____

4. If the rug were 5 units long and 4 units wide, how much greater would its area be than the one shown in the model?

   ________ square units greater

Solve.

5. Shawna wants to cover her living room with wall-to-wall carpeting. The carpet she wants is $24 for each square yard. If her living room is 3 yards long and 5 yards wide, how much will the carpeting cost? Use a calculator.

   ____

6. Kaitlin wants to hang a large painting on her wall. She wants the painting to be square in shape and have an area of at least 32 square feet. What is the smallest area her painting can have? Explain.

   ____

24.7 Explore Volume

Use the diagram to solve.

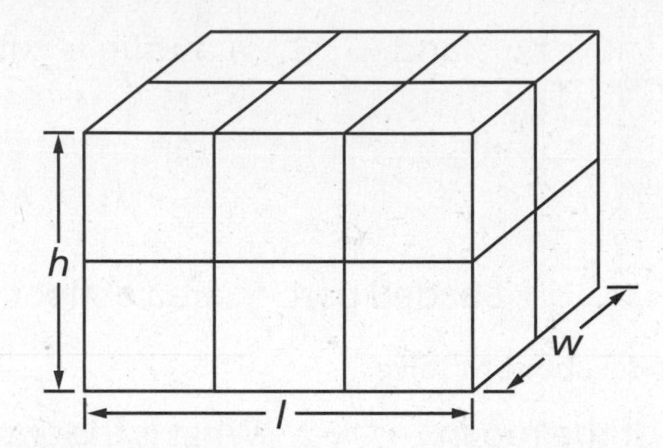

1. How many cubic units long is the box? ________ cubic units

   How many cubic units wide? ________ cubic units

   How many cubic units high? ________ cubic units

2. What is the volume of the box in cubic units?

   ________ cubic units

Solve. Use connecting cubes to help.

3. Miko's jewelry box is 4 units long, 5 units wide, and 6 units high. What is its volume?

   ________ cubic units

4. Haki built a box out of 24 cubic units. What could the dimensions of the box be?

   ____

Solve.

5. A ring box is 3 inches long, 2 inches wide, and 1 inch high. What is the volume of 2 of these boxes?

   ________ cubic inches

6. Each block is about 1 cubic inch in volume. About how many blocks could you fit into a box that is 4 in. long, 4 in. wide, and 4 in. high?

   about ________ blocks

Parts of a Whole

Solve.

1. One half of the wall is blue and one half is yellow. What fraction shows the part of the wall that is blue?

____

2. A muffin is cut up into 3 equal parts. One of the parts has been eaten. What fraction of the muffin has been eaten?

____

Solve.

3. Betty baked a meatloaf. She cut it into 5 equal slices. The family ate 3 of the slices. What fraction of the meatloaf did they eat?

____

4. Tom baked an apple pie and cut it into equal pieces. Tom ate one piece, which was $\frac{1}{6}$ of the pie. How many pieces did he cut the pie into?

____

Solve.

5. Tony's Pizzeria cuts their 8-inch pizzas into 4 equal slices. Martelli's Pizzeria cuts their 8-inch pizzas into 6 equal slices. Andre had a slice of pizza at both pizzerias. At which pizzeria did Andre eat more pizza? Explain.

____

6. A loaf of bread is cut into 8 equal slices. How much of the bread is left after 6 slices have been used for sandwiches?

____

25.2 Explore Equivalent Fractions

Solve. Use fraction models if you need help.

1. Lenny colored $\frac{1}{2}$ of his picture. What is another fraction that tells the part of the picture he colored?

   ____

2. A painter has painted $\frac{2}{8}$ of a ceiling. What is an equivalent fraction for this?

   ____

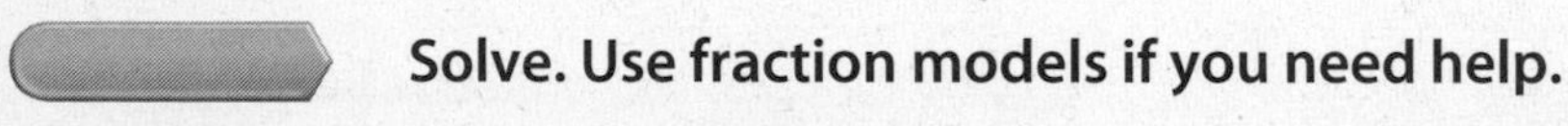

Solve. Use fraction models if you need help.

3. Phillip has a box that is divided into 4 equal sections. He fills 2 of the sections with sand. Write two equivalent fractions that tell how much of the box is filled.

   ____

4. A granola bar is cut into 3 equal parts. Grace eats one part. Write two equivalent fractions that tell how much of the bar she ate.

   ____

Solve. Use fraction models if you need help.

5. A circular tablecloth has 8 equal sections. Two sections are white, two are red, two are blue, and two are black. What part of the tablecloth is not white?

   ____

   What is another fraction you can use to name this part?

   ____

6. A spinner is divided into 6 equal sections. The sections are numbered in order from 1 through 6. What part of the spinner has even numbers on it?

   ____

   What is another fraction you can use to name this part?

   ____

Fractions in Simplest Form

Solve.

1. Ally divides a rectangle into 4 equal parts. She colors 2 of the parts green. What part of the rectangle is green?

   ____

   How can you write the fraction in simplest form?

   ____

2. A 6-inch ruler breaks off at the 2-inch mark. What part of the ruler broke off?

   ____

   How can you write the fraction in simplest form?

   ____

Solve.

3. A floor is being covered with 16 tiles, each the same size. If 8 of the tiles have been laid down, in simplest form, what fraction of the floor is covered?

   ____

4. Jeanette cuts an apple into 10 pieces. She eats 4 of them. In simplest form tell what fraction of apple Jeanette ate.

   ____

Solve.

5. A cake is cut into 12 equal pieces. So far, children have eaten 4 pieces of the cake. What fraction of the cake was eaten?

   ____

   How can you write the fraction in simplest form?

   ____

6. Meg designed her book cover by dividing it into 16 equal squares. She drew stripes on 5 of the squares and polka dots on 7 squares. In simplest form, what fraction of the book cover has neither stripes nor polka dots on it?

   ____

Compare and Order Fractions

Solve.

1. Pete and Sal share a bag of chips. Pete eats $\frac{1}{4}$ of the chips and Sal eats $\frac{3}{4}$ of the chips. Who eats more?

   ____

2. If $\frac{2}{5}$ of the class are boys and $\frac{3}{5}$ are girls, are there more boys or girls?

   ____

Solve.

3. It takes $\frac{3}{4}$ of an hour for Randy to walk from home to school. It takes $\frac{1}{2}$ hour for him to walk from home to the mall. Does Randy live closer to school or to the mall?

   ____

4. Alice has finished $\frac{2}{3}$ of her homework. Sam has finished $\frac{1}{2}$ of his homework. Who has more homework left to do, Alice or Sam?

   ____

Solve.

5. In a recipe for a fruit salad, Marta adds $\frac{1}{2}$ pound of apples, $\frac{3}{4}$ pound of grapes, and $\frac{1}{3}$ pound of cherries. How can you list the ingredients in order from greatest to least amount added?

   ____

6. Jack and Sandra each have \$100 in savings. Jack spent $\frac{1}{2}$ of his savings on a new coat and $\frac{3}{8}$ of the savings on a new pair of sneakers. Sandra spent $\frac{2}{5}$ of her savings on a new coat. Who spent more money on a coat?

   ____

Parts of a Group

Solve.

1. A box of crayons has 3 red crayons and 4 blue crayons. What fraction of the crayons is red?

____

2. Sheila has 3 dimes and 2 nickels. What fraction of the coins are nickels?

____

Solve.

3. Michelle exercises on Mondays, Wednesdays, and Fridays each week. What fraction of the days of the week does she exercise?

____

4. Elizabeth writes each letter of her first name on separate index cards. What fraction of the cards have vowels?

____

Solve.

5. A quilt has 4 rows of red squares and 6 rows of white squares. Each row has 5 squares. In simplest form, what fraction of squares is not red?

____

6. A jar has 12 red, 8 blue, and 10 white marbles in it. In simplest form, what fraction of the marbles are either red or blue?

____

Explore Finding Parts of a Group

Use counters to solve.

1. If you color $\frac{1}{3}$ of 6 circles blue, how many circles will you color blue?

   ______ circles

2. If you take away $\frac{1}{4}$ of 8 counters, how many counters will you take away?

   ______ counters

Solve. Use counters to help.

3. There are 12 children at a lunch table. If $\frac{2}{3}$ of them have pizza for lunch, how many children at the table have pizza?

   ______ children have pizza

4. The winter season lasts for about $\frac{1}{4}$ of the year. For about how many months does winter last?

   ______ months

Solve. Use counters if you need help.

5. Tina reads for $\frac{3}{4}$ of an hour. Mike reads for $\frac{1}{4}$ of an hour. For how many more minutes does Tina read than Mike?

   ______ minutes more

6. There are 18 coins in a drawer. If $\frac{1}{6}$ of the coins are pennies, $\frac{2}{6}$ are nickels, and the rest are dimes, how many are dimes?

   ______ dimes

   In simplest form, what fraction of the coins are dimes?

   ____

Problem Solving: Skill
Check for Reasonableness

Solve. Check for reasonableness.

1. In the Walsh family $\frac{1}{4}$ of the 8 children are girls. How many children are girls?

____

2. Marco gave his dog $\frac{1}{3}$ of the 9 biscuits left. How many biscuits did Marco give his dog?

____

Solve. Check for reasonableness.

3. A pound cake is cut into 10 equal slices. If $\frac{3}{5}$ of the slices were served for dinner, how many slices were served?

____

4. Karen's school day starts at 8:00 A.M. and ends at 2:00 P.M. Today she has a $\frac{1}{2}$ day. What time does her school day end?

____

Solve. Check for reasonableness.

5. Gary goes to school for $\frac{8}{24}$ of the day. He spends $\frac{4}{24}$ of his day with friends. He does his homework for $\frac{2}{24}$ of the day, and he also practices piano for $\frac{2}{24}$ of the day. He sleeps for the rest of the day. For how many hours does Gary sleep each day? Explain.

____

____

6. There are 15 counters on the desk. If $\frac{3}{5}$ of the counters are yellow, $\frac{1}{5}$ are white, and the rest are black, how many are black?

____

____

26.1 Mixed Numbers

 Solve.

1. The model shows how Ann colored two circles. What mixed number shows how many circles she colored?

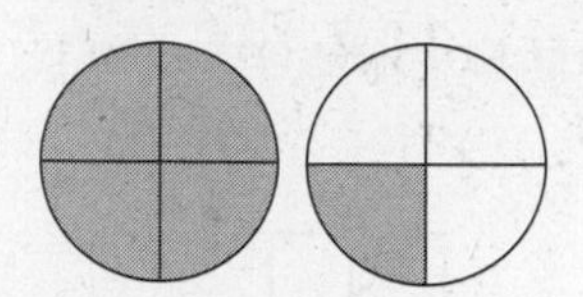

____

2. The Winston family ate 1 whole pizza and $\frac{2}{3}$ of another. How much pizza did the family eat?

____

 Solve.

3. The picture shows how much flour Anne put into a cake recipe. How many cups of flour did she use?

____

4. Rod shaded $3\frac{3}{4}$ circles on a math page. Draw a model to show how he colored them in. Use a separate sheet of paper.

____

 Solve.

5. Mark wants to decorate a square frame by gluing crayons onto it. The frame is 5 in. × 5 in. How long is the crayon below to the nearest $\frac{1}{2}$ inch? ________ inches
How many crayons will Mark need to cover his frame?
_____ crayons

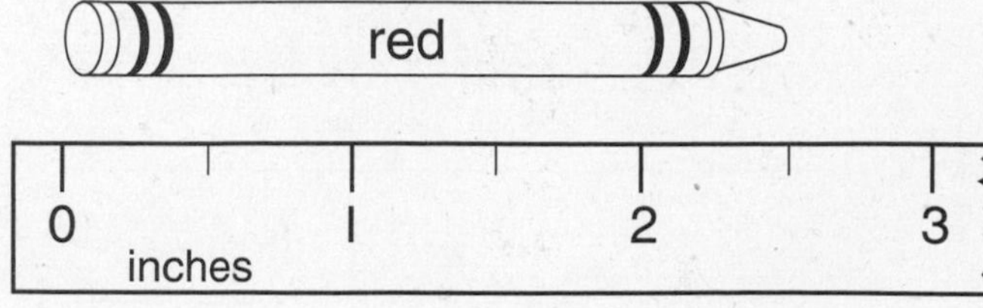

6. Seth and Holly draw models to show how far each walked. Seth draws two circles and divides them into thirds. He shades the entire first circle, and one section of the second. Holly draws two circles, but she divides them in halves. She colors the entire first circle, and one section of her second circle.
If each circle is equal to a mile, how far did Seth walk? ________ mi
How far did Holly walk? ________ mi
Who walked farther? ________

Explore Adding Fractions

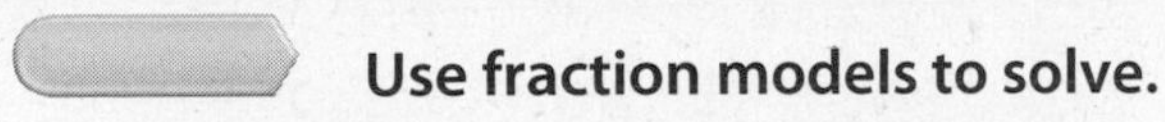

Use fraction models to solve.

1. Kitty the cat ate $\frac{1}{6}$ of her food in the morning and $\frac{2}{6}$ in the afternoon. How much of her food did Kitty eat so far?

$\frac{1}{6}$

$\frac{1}{6}$	$\frac{1}{6}$

____

2. A plant grew $\frac{1}{5}$ of an inch during the first week and $\frac{3}{5}$ of an inch the next week. How much did the plant grow in the two weeks?

$\frac{1}{5}$

$\frac{1}{5}$	$\frac{1}{5}$	$\frac{1}{5}$

____

Solve. Use fraction models if you need help.

3. A model car storage box is divided into 8 equal sections. $\frac{3}{8}$ of the sections have model sports cars and $\frac{2}{8}$ have model trucks. The rest of the box is empty. What fraction of the box is filled so far?

____

4. Of the dozen eggs in a box, $\frac{4}{12}$ have been colored pink and $\frac{3}{12}$ have been colored blue. The other eggs have not been colored. What fraction of the eggs have been colored?

____

Solve. Use fraction models if you need help.

5. Ricky paints $\frac{1}{8}$ of his room on Monday, $\frac{3}{8}$ on Tuesday, and $\frac{2}{8}$ on Wednesday. What fraction of his room did Ricky paint by the end of the day on Wednesday?

____

6. There are 12 apples in a bag. Everett eats $\frac{1}{6}$ of them and Lilly eats $\frac{2}{6}$ of them. What fraction of the apples have been eaten so far?

____

How many apples are left in the bag?

____

Explore Subtracting Fractions

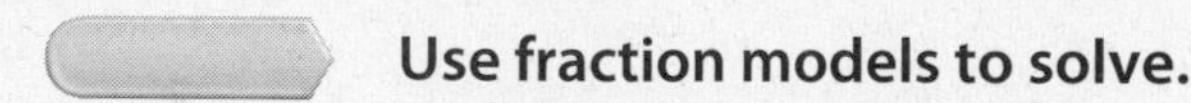

Use fraction models to solve.

1. Lorrie found $\frac{3}{4}$ of an apple pie in the refrigerator. She ate $\frac{2}{4}$ of the original pie. What fraction of the pie was left?

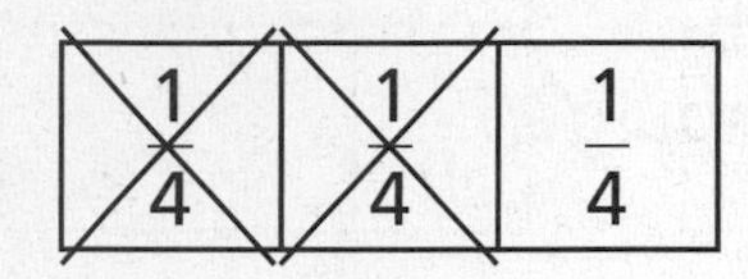

_______ of the pie

2. There is $\frac{7}{8}$ of a quart of milk in a bottle. Brianne pours $\frac{2}{8}$ of a quart of milk into a glass. How much milk is left in the bottle?

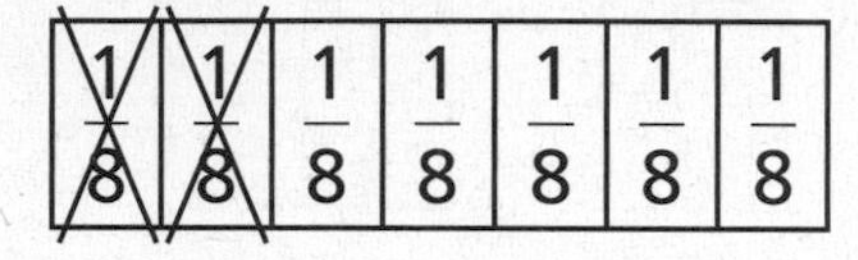

____

Solve. Use fraction models if you need help.

3. A box of crayons fell on the floor. If $\frac{7}{12}$ of the crayons fell out, what fraction of the crayons are still in the box?

_______ are still in the box

4. Alex ran $\frac{8}{12}$ of a mile. Rhea ran $\frac{5}{12}$ of a mile. In simplest form, how much farther did Alex run than Rhea?

_______ of a mile farther

Solve. Use fraction models if you need help.

5. Ben found $\frac{7}{10}$ of a pound of flour in the pantry. He needed to use $\frac{3}{10}$ of a pound of flour for cupcakes that he was baking. His mom said that she needed $\frac{3}{10}$ of a pound of flour for dinner. After Ben and his mom use the flour, how much will be left in the bag?

_______ of a pound

6. Mea baked a peach pie and an apple pie. At the end of the day, $\frac{7}{8}$ of the peach pie was left and $\frac{3}{8}$ of the apple pie was left. In simplest form, how much more peach pie was left than apple pie?

_______ of a pie

Add and Subtract Fractions

Solve.

1. If $\frac{1}{5}$ of Jerell's postcard collection is from Europe, and $\frac{3}{5}$ are from the United States, what part of the collection is from Europe or the United States?

   ________ of the collection

2. Mitchell added $\frac{2}{3}$ of a teaspoon of salt to his soup. Later he added another $\frac{1}{3}$ teaspoon of salt to his hamburger. How much more salt did he add to his soup than his hamburger?

   ________ of a teaspoon more

Solve.

3. Harry has a great CD collection. $\frac{5}{9}$ of his collection is rock music. $\frac{2}{9}$ of the collection is country music. In simplest form, how much more of the collection is rock music than country music?

   ________ more of the collection

4. Boris is making clay action figures. For each one, he uses $\frac{2}{6}$ of a pound of clay. If he makes 2 action figures, how much clay does Boris use? Write the answer in simplest form.

   ________ of a pound of clay

Solve.

5. Mr. Harvey has 32 third-grade students. They tested the flavor of a new bubble gum. $\frac{17}{32}$ liked the gum and $\frac{7}{32}$ students could not decide. The rest did not like the flavor. In simplest form, what fraction of the class did not like the flavor of the gum?

   ________ did not like the flavor

6. Manuel has $\frac{11}{12}$ of a pound of chopped meat. He uses $\frac{3}{12}$ of a pound to make a hamburger. In simplest form, how much chopped meat will he have left if he makes 3 hamburgers?

   ________ of a pound

Probability

Solve. Use the words *certain, likely, unlikely,* or *impossible.*

A number cube has 6 sides numbered 1 through 6.

1. What is the probability that you will toss a 7 if you toss the number cube?

____

2. What is the probability that you will toss a 6 if you toss the number cube?

____

Solve. Use the words *certain, likely, unlikely,* or *impossible.*

A bag has 6 green grapes and 8 red grapes.

3. Keisha is going to pick one item from the bag. What is the probability that she will pick a grape?

____

4. What is the probability that she will pick a red grape?

____

Solve. Use the words *certain, likely, unlikely,* or *impossible.*

A spinner is divided into 10 equal sections numbered 1 through 10.

5. If Clea spins the spinner, what is the probability that she will land on a number less than 3?

____

6. What is the probability that he will land on either an odd or even number?

____

Explore Finding Outcomes

 **Solve.**

1. Sylvia tosses a red and yellow counter 6 times. What are the possible outcomes? How many times do you predict the counter will land on red?

   ____

2. Mick has a number cube. Three sides are green. Two sides are red. One side is blue. Mike tosses the cube 3 times. What are the possible outcomes? How many times do you predict the cube will land on green?

   ____

 **Solve.**

Julia plays a board game that has 2 red, 4 blue, and 4 yellow slips of paper in a bag. To play the game Julia picks a slip of paper, and then moves to that color on the board. Then she puts the paper back in the bag and does it again. She does this 10 times all together.

3. What are the possible outcomes?

   ____

4. Do you think Julia will choose more yellow, blue, or red slips of paper? Explain. ______________________

   ____

 **Solve.**

5. Theresa has 3 pairs of black shoes, 2 pairs of brown shoes, and 1 pair of sneakers. She does not have a light in her closet. She is looking for a pair of black shoes. She reaches into her closet and pulls out shoes one pair at a time.

   What color is she most likely to pick? ______________________

   What color will she most likely pick the second time?

   ____

6. Ashton has 4 brown hamsters, 3 spotted hamsters, and 1 white hamster. If he reaches in without looking to pet one, which color hamster is he mostly likely to pet?

   ____

   Predict which he is least likely to pet. ______________________

   If Ashton picks up a brown hamster, and reaches in again, predict which color he will pet this time.

   ____

26.7 Problem Solving: Strategy
Make an Organized List

Solve. Use an organized list.

1. You have a red shirt and a white shirt and 1 pair of blue pants. How many different outfits can you make?

_______ different outfits

2. For lunch Kevin can have either ham or turkey on white or wheat bread. How many different kinds of sandwiches does he have to choose from? What are they?

____

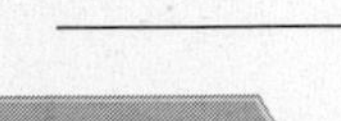

Solve. Use an organized list.

3. Ezra will put a pen, pencil, and eraser in his pencil case. He can choose from a black or blue pen and a wooden or plastic pencil. He only has one eraser. How many different combinations can he choose from?

_______ different combinations

4. The teacher will choose 3 out of 4 students to form a reading group. How many different ways can she form the group?

____

Solve. Use an organized list.

Here is the dinner menu at Old Diner Inn.

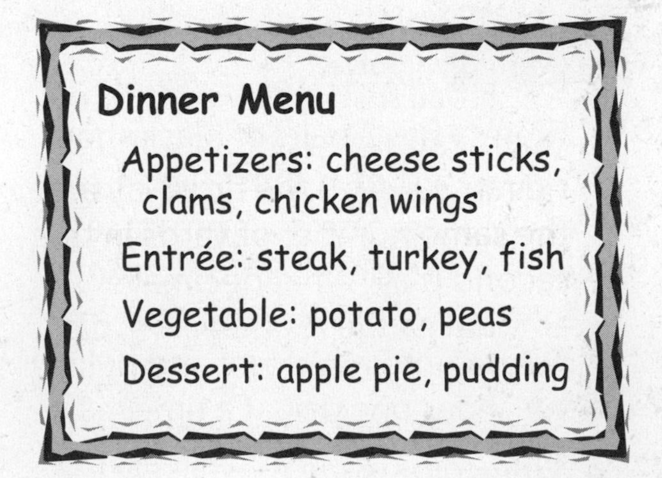

5. Emily will choose an entrée, vegetable, and dessert, but no appetizer. How many different choices does she have?

____

6. Janie is very hungry. She orders an appetizer, entrée, vegetable, and a dessert. How many different choices does she have?

____

Explore Fractions and Decimals

27.1

Solve.

1. Kurt has a dollar. He bought a pack of gum for 50 cents. What fraction of the dollar did he spend?

____

2. Amanda must paint a floor with 100 equal-sized squares. She has painted 20 of them. Write a decimal for the part of the floor that has been painted.

____

Solve.

3. The Marteen ranch has 100 cows. Yesterday, 45 of them escaped from the barn. What decimal amount of the cows escaped? How can you write the decimal amount as a fraction?

____

4. Wu has 100 model cars. If $\frac{20}{100}$ of them are broken, how many are broken?

____

Solve.

5. A football field is 100 yards long. Abdul ran two plays in football. He ran 30 yards the first play and the same number of yards in the second play. What decimal amount of the football field did he run?

____

How can you write that amount as a fraction?

____

6. Milton ran six-tenths of a mile in gym. Lisa ran $\frac{60}{100}$ of a mile. Lisa says that she ran farther than Milton. Is Lisa correct? ____________

Explain.

____

____

27.2 Fractions and Decimals

Solve.

1. Elway spent 90¢ on a drink. How can you write this amount as a fraction?

____

2. Martin's guinea pig weighs $\frac{6}{10}$ of a pound. Write this weight as a decimal number.

________ of a pound

Solve.

3. Byron gave his little sister $\frac{5}{10}$ of a dollar to buy a newspaper. How much did he give her?

________ cents

4. There are 100 cartons of milk in the school cafeteria refrigerator. Dennis moved 45 cartons before lunch. What part of the milk in the cafeteria did he move? Write your answer as a decimal.

____

Solve.

5. Nellie needs a piece of string that is 0.8 of a meter. She has found a piece that measures 85 centimeters in length. Is this piece long enough? Explain.

____

____

6. Leroy wants to buy a bus pass for $1.50. He has 4 quarters and 2 dimes in his left pocket and 7 nickels in the right one. What decimal part of a dollar will he have left over?

____

Decimals Greater Than One

Solve.

1. Raymond has $6\frac{25}{100}$ boxes filled with newspapers to recycle. Write this amount as a decimal.

_________ boxes

2. Sharika has a garden snake. It is $1\frac{5}{10}$ feet long. How would you write the number of feet as a decimal?

_________ feet

Solve.

3. Stan ran $1\frac{5}{100}$ of a mile. How can you show the distance as a decimal number?

____

4. Lorrie collects stickers. She has filled 3 full pages and $\frac{3}{10}$ of another page of her album. Write the number of pages filled as a decimal number.

_________ pages

Solve.

5. Gus has two small boxes of chocolate. Each box holds 10 pieces. One box is filled. The other is $\frac{4}{10}$ filled. Write the amount of boxes of chocolate he has as a decimal.

_________ boxes

6. Ralph worked on his science project for 3.8 hours. Sara worked on hers for $3\frac{80}{100}$ hours. Ralph said he worked longer. Sara said she worked longer. Who is correct?

____

Explain.

____

Compare and Order Decimals

Solve.

1. The third-grade children held a contest to see who could run farthest in a certain amount of time. Nadia ran 1.16 miles. Justine ran 1.21 miles. Who ran farther?

____

2. The Thomas children saved money for a birthday gift for their mom. Vera saved \$2.49. Emma saved \$3.49. Gill saved \$3.21. List the children in order from the one who saved the most to the one who saved the least.

____

Solve.

3. Every three months Mr. Stone measures the height of his daughters Rachael and Esther. The last time Mr. Stone measured the girls, Rachael had grown 1.36 inches and Esther had grown 1.63 inches. Which girl grew less?

____

4. Andrea, Nicholas, and Louis drew straws to decide their batting order on the baseball team. The one with the longest straw bats first. Andrea drew a straw that was 3.63 inches long. Nicholas's straw was 2.64 inches long. Louis's straw was 3.7 inches long. In what order will the players come to bat?

____

Solve.

5. Grandma Midge's cookie recipe calls for 2.5 cups of flour. Arlis added $2\frac{7}{10}$ cups of flour to his recipe. Mitchell added $2\frac{3}{10}$. Brooke added $2\frac{50}{100}$. Who put too much flour in the recipe? Who added the right amount?

____

6. Greg and Tina went to the mall. Greg had \$4.38 and he spent \$2. Tina had \$3.40 and she spent \$1. Which person had more money left? Explain.

____

Problem Solving: Skill
Choose an Operation

Solve. Tell how you chose the operation.

1. A pack of batteries costs $4.75 in the hardware store. The same pack costs $3.50 in the office supply store. How much more are the batteries in the hardware store?

____

2. Mr. Maxwell bought a hammer for $5.75 and a wrench for $4.20. How much did he spend in all?

____

Solve. Tell how you chose the operation.

3. Clarise has $6.50. She owes her brother $2.25. How much will Clarise have if she pays her brother the money she owes him?

____

4. Admission to the zoo is $12.95 for adults. Admission to the amusement park is $15.75 more than admission to the zoo. How much is admission to the amusement part?

____

Solve. Tell how you chose the operation.

5. Renee started with $25 in her checking account. She deposited another $8.85, then paid her grocery bill for $12.65. How much does she left after buying a new CD for $15? Explain.

____

6. A paperback book costs $0.95 more than a magazine and a magazine costs $0.85 more than a puzzle. If the puzzle costs $3.45, how much do the book and magazine cost all together?

____

28.1 Explore Adding Decimals

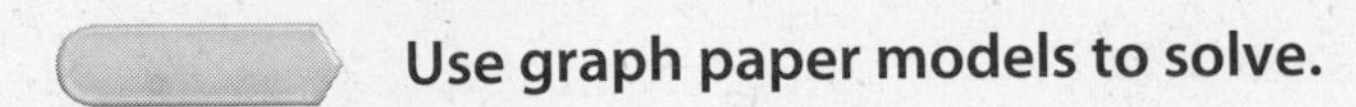

Use graph paper models to solve.

1. Mel swims 1.4 kilometers during practice and 1.3 kilometers after practice. How far does Mel swim in all?

_______ kilometers

2. On Monday 1.25 inches of rain fell. On Tuesday 1.25 inches of rain fell. How many inches of rain fell in all?

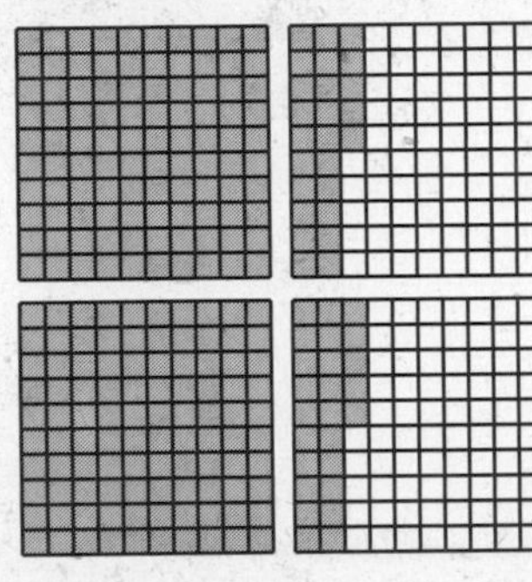

_______ inches

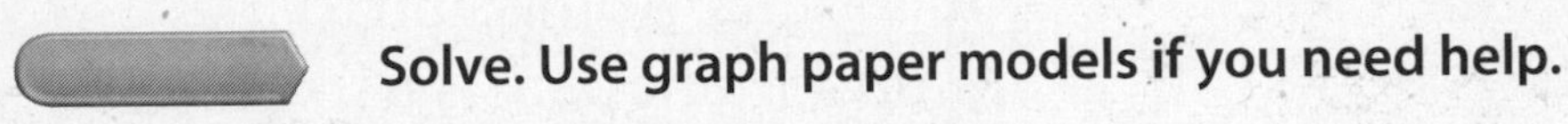

Solve. Use graph paper models if you need help.

3. Mark's dad is putting up a tile wall in the bathroom. He puts up 2.6 rows of tile, then takes a break. He puts another 3.5 rows of tile before lunch. How many rows of tile are up so far?

_______ rows of tile

4. Petra and Jana are putting sugar into a bowl to make fudge. Petra has put 0.5 of a cup in the bowl. Jana has put in 0.75 of a cup. How much sugar is in the bowl?

_______ cups

Solve. Use graph paper models if you wish.

5. A blizzard hit the town of Springfield. For two hours it snowed at the rate of 2.68 inches an hour. The third hour it snowed 3.31 inches. How much did it snow in all three hours?

_______ inches

6. When he left for work, Mr. Jenson's car odometer read 5.6 miles. Mr. Jenson drove 2.75 miles to work and the same distance back home. Now what does the odometer read?

_______ miles

Add Decimals

Solve.

1. Tommy is growing a tomato plant. It grew 1.32 cm last week. This week it grew 1.14 cm. How much did it grow in two weeks?

   ________ centimeters

2. Brett has $1.30 in his pocket. He has $1.60 in his room at home. How much money does he have in all?

   ____

Solve.

3. Holly's pencil has a mass of 9.65 grams. What is the mass of two of these pencils?

   ________ grams

4. Dana rode her bike 2.38 kilometers to the bakery. She bought some doughnuts and rode 1.35 kilometers to her grandma's house. How many kilometers did she ride in all?

   ________ kilometers

Solve.

5. Arlene and three of her friends went to the movies. Each ticket to the movie was $7.25. They spent a total of $5.50 on popcorn. How much did they spend in all?

   ____

6. Jeff has two hamsters. Each hamster weighs 4.8 ounces. His guinea pig weighs 8.12 ounces. If you put all three animals on a scale, what weight will the scale read?

   ________ ounces

Problem Solving: Strategy Solve a Simpler Problem

Solve.

1. Sven carried three stones to the front of his house. Two stones had a mass of 9.4 kilograms each. The other had a mass of 8.75 kilograms. What is the total mass of the stones?

   ________ kilograms

2. Mario and Cindy rode a camel at the county fair. The camel ride was 0.6 miles long. Mario rode once. Cindy rode twice. What is the total distance for all of their rides?

   ________ miles

Solve.

3. Nick watered the lawn for two hours. He used 20.25 liters of water each hour. He then washed two loads of clothes. Each wash used 6.35 liters of water. How many liters of water did Nick use to water the lawn and wash clothes?

   ________ liters

4. The Carlsons went to see a basketball game. There are 4 children and 2 adults in the family. The children's tickets were $4.25 each. The adult tickets were $9.75 each. What was the total price of the tickets?

   ____

Solve.

5. Grace has two 50-foot rolls of tape. She will use it on three projects. The first project will take 21.4 feet of tape. The second project will use 18.24 feet. The third project will use 12.36 feet of tape. How many feet of tape will she have left?

   ________ feet left

6. Trisha took her cat, Butterball, to the vet. Butterball had an exam that cost $25.97. She had her nails clipped for $15.50. Trisha also bought 2 cans of diet cat food. Each can cost $1.50. How much did Trisha spend in all at the vet?

   ____

Explore Subtracting Decimals

Use graph paper models to solve.

1. A salt shaker has 1.7 ounces of salt left. A pepper shaker has 0.6 ounces left. How much more salt is left than pepper?

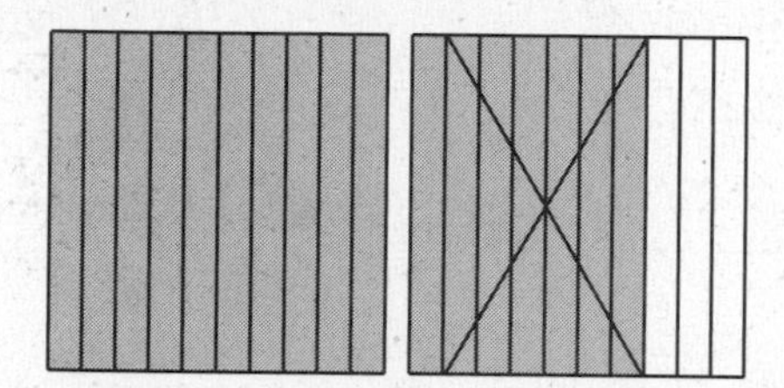

_______ ounces more

2. Jack lives 2.6 miles from school. Ashley lives 1.2 miles from school. How much farther away from school does Jack live than Ashley?

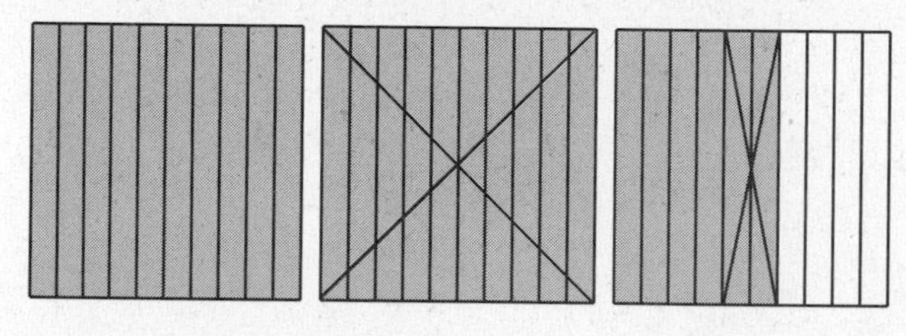

_______ miles farther away

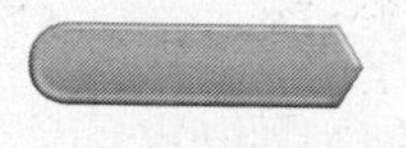

Solve. Use graph paper models to help.

3. A bag of grapes weighs 2.75 pounds. A bag of cherries weighs 3.55 pounds. How much more does the bag of cherries weigh than the grapes?

_______ pound more

4. A container can hold 2.5 quarts of liquid. If you pour a 3.2 quart bottle of water to the container until it is full, how much water is left in the bottle?

_______ quart

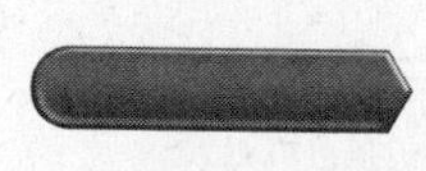

Solve. Use models if you wish.

5. Manuel's dog needs to eat 3 kg of dog food each day. Manuel feeds him 1.8 kg of food in the morning, 0.75 kg in the afternoon and the rest in the evening. How much food does he feed the dog in the evening?

_______ kilograms

6. In a box of animal crackers, 0.25 of all the crackers are camels. What decimal amount represents all the other animal crackers in the box?

____

28.5 Subtract Decimals

Solve.

1. George has $5.50. He spent $2.47 for a hamburger and fries. How much does he have left?

____

2. One fish tank holds 8.2 gallons of water. Another can hold 6.5 gallons. How much more water can the larger tank hold?

____

Solve.

3. King-size drinks at Twin Cinema hold 19.45 ounces. King-size drinks at the Middlebrook Cinema hold 21.35 ounces. How many more ounces are the king-size drinks at the Middlebrook Cinema?

_______ ounces more

4. Jody's dad is painting their house. He bought a 5-gallon container of paint. While opening the container, he spilled 1.87 gallons. How much paint did he have left?

_______ gallons left

Solve.

5. Trudi has 58.5 m of wood molding for her room. Each of 2 sides of her room will use 8.25 m of molding. The other two sides will use 6.15 m each. How many meters of molding will she have left over?

_______ m left

6. Judy has three kittens named Wilbur, Banana, and Addison. The total weight of the three kittens is 3.56 kilograms. Wilbur weighs 1.23 kilograms. Banana weighs 1.11 kilograms. How much does Addison weigh?

_______ kilograms